ENTANGLED STATES

ENTANGLED STATES

A LIFE ACCORDING TO QUANTUM PHYSICS

KARMELA PADAVIC-CALLAGHAN

BEACON PRESS, BOSTON

BEACON PRESS
24 Farnsworth Street
Boston, Massachusetts
www.beacon.org

Beacon Press books
are published under the auspices of
the Unitarian Universalist Association of Congregations.

Printed in the United States of America

29 28 27 26 8 7 6 5 4 3 2 1

This book is printed on acid-free paper that meets the uncoated paper
ANSI/NISO specifications for permanence as revised in 1992.

Text design and composition by Kim Arney

*Library of Congress Cataloguing-in-Publication
Data is available for this title.*
Hardcover ISBN: 978-0-8070-1698-5
E-book ISBN: 978-0-8070-1700-5
Audiobook: 978-0-8070-2334-1

The authorized representative in the EU for product safety and
compliance is Easy Access System Europe 16879218, Mustamäe tee 50,
10621 Tallinn, Estonia: https://beacon.org/eu-contact.

To Bennett, Alex, Jo,
and everyone else who loved me lately

CONTENTS

INTRODUCTION

I took my first physics class in the seventh grade. It was taught by one of the most feared teachers in the school, a short man with a prominent moustache who wore the kind of chunky white sneakers that belong to fashion girls now but in 2003 were solely the choice of middle-aged dads. He was not a good teacher. He assumed that everyone was always cheating, he lectured robotically, and he never led fun experiments such as turning a potato into a battery. When he joined my father's amateur bocce ball league, despite all his haughtiness in the classroom, he could not understand how a tournament gets split into brackets. My dad and I laughed about it. His class, however, changed my life.

In that class, the teacher claimed, matter-of-factly, I would learn how to uncover the truth about reality through words and equations—that physics could teach me how to build a model of our world piece by piece. When the teacher spoke, there was no poetry in his voice, nor joy. But when he started scribbling on the blackboard, between his chalky lines I saw the glimmers of something that would put its hooks in me forever.

The history of physics is long, stretching from ancient scholars deriving the circumference of the Earth with sticks to modern governments pouring billions of dollars into particle colliders. Somewhere in between, though, the euphoric feeling of discovery that is at the discipline's core—the deeply human feeling of solving a mystery on its own terms and timeline—started to dim, maybe even fade.

I devoted two decades to the study and practice of physics, then failed to land the kind of research job I would have needed to establish a career. I left the formal world of science with a slightly broken heart. But physics refused to leave me. My seventh-grade dream of becoming a physicist had dissolved, but in its aftermath I discovered that a physicist's ways of seeing were now solidly embedded in my mind. When I had first set out to understand things like what gravity is and how electricity works, I hadn't anticipated becoming so embroiled in meaning-making—in feeling so much for the world as a consequence of trying to capture it with mathematics. No one had really told me about that part: the never-ending pools of wonder, awe, and significance that I am now somehow always swimming in, helping me to make sense of myself and my world. With this book, I want to tell you about it.

I have always wanted to know everything, and pursuing physics seemed like a safe bet for getting there. I organized my life around it. At sixteen, I moved across the world by myself to attend schools that would prepare me to become a physicist. This felt necessary, even unavoidable. The promise of knowing enticed me that much. And it paid off—just not in the form of a job that requires a lab coat or a classroom lectern. When I am asked why I left physics if I loved it so much, I shamelessly assert that I was a bad physicist. Yet, my brain is still always on physics—I never really managed to come down from that high.

Physics offers itself to me when I try to make sense of all the paths my life did and did not take. It reassures me when I try to reconcile and cobble together all the identities that I feel describe me, and it is there when I am waging war against my body or distrusting my memories, providing anchors and hardpoints for my thinking. Sometimes it comforts me. Sometimes, when it connects my life experience to what seems to be the deep truth of the physical universe, it sets my mind on fire. At its best, it makes me feel like I am ready to make sense of whatever might happen to me next, be it good or bad. I hope this book can give you a chance to find that connection too, and let it rock you a little—sometimes soothingly, sometimes right off your feet.

Physics, as I have been taught to think about it, is the science of how the world is and why it is that way. When I was a high school teacher, I would write on the blackboard that physicists study matter, energy, and how they all interact. But here on Earth, matter, and interacting matter, is everything, everywhere, all the time. So, I am offering you a book about physics as I have learned to live with it. In other words, a book about everything.

For several years now, I have been a full-time physics reporter, which means that I spend my days talking to working physicists and learning the whats and whys and hows of their work. Our conversations overflow with facts, definitions, and technical details. Then, I sit at my desk and summarize it all for the science-curious reader. In this book, I want to engage you in a different way. I want you to imagine chatting with me at a party where I brim with excitement, not because I am trying to show off facts, but because I am making sense of being alive, one day after another. And you certainly don't have to be a "science person" to care about that.

I wish I could say that finding new ways to look at life is as simple as me telling you that there are trillions of particles passing through you right now, or that empty space isn't actually empty, but these are just fun facts. I share them to impress family members, or in restless moments with someone's child. But I want this book to give you something not just to remember, but that you can feel and ruminate on.

When I say this is a book about everything, I don't mean everything that you can look at under the microscope, or everything that you can calculate with an equation. I mean everything like a bad tooth that almost killed me once, like tiny objects for whom cause and effect don't always follow the usual order, like how an atom can be cooled with a laser and memories can be built from dirty diamonds, like the fear of letting people get to really know you, like the connection between music and the nature of time, like the looming quantum multiverse, like why I love black eyeliner so much. I mean that I want to tell you a story about everything that has ever happened to me: a series of events that cannot be disconnected from everything that has ever happened to every particle and every atom of our physical world.

Several summers ago, I was at the pool with a friend, trying to tell them a story about quantum computers. I often report on quantum computing, so my story was well-rehearsed:

Conventional computers operate on 1s and 0s, the kind that you saw in the '90s as neon green columns in *The Matrix*: each 1 means that a tiny switch somewhere in the computer gets flipped to "on," while a 0 turns it "off." Quantum computers—which rely on effects that only exist for very small or very cold objects, which we then call "quantum"—offer more than two choices: more states of being than fully "on" or fully "off." They can mix their "on" states and "off" states, combinations of 1s and 0s that have no analogue in more conventional devices. This greatly contributes to their computational power, making them capable of processing a lot more information. To run certain algorithms, for example, a machine with only 10 quantum bits could replace a conventional computer with 1,024 bits. Building a fully practical quantum computer has proven extremely difficult, but scientists take this as reason to hope that quantum computers could eventually solve problems—from answering open questions in fundamental physics to designing better solar panels and drugs—that are currently completely intractable, even for our most advanced supercomputers.

My friend squinted in the sun, busied their hands with a crumpled food magazine. They were waiting for me to say more: something conveying more meaning than the assertion that some new machine will eventually do some amorphously big thing. So, I told them the part of the story that never makes it into my articles.

I've never been able to pick a single way of being, I told them, and this has always troubled me. I was an odd child and often a sad one, probably because I never knew how to fit the chaotic expanse of my interior world into the shape of an easily classified, easily digestible person. I grew up to be a scientist but also a writer, to be bisexual and nonbinary, to run five miles a day but also bake a three-layer cake most weeks, to understand vector calculus but also love friends who dabble in magic (to me, both are beautiful). As I wrote article after article about quantum computers, and heard physicists say again and again that some new device was only useful and interesting because its components could be put in a combination of two seemingly contradictory states at once, I began to wonder why I am so bothered by my own inability to be easily classified. I had to admit to myself that it's sort of silly to be upset about a quality of mine that is not actually out of place in the physical world.

When I stopped talking, my friend cracked a smile underneath their candy-colored bucket hat, followed by a soft "Hell yeah." Several weeks later I was in their kitchen at a small soup-themed dinner party, filling my bowl and talking about X-ray lasers. I had off-handedly mentioned writing about them, which unexpectedly invited many enthusiastic questions from other soup enthusiasts. No one in that room cared much about the electron accelerators or fine-tuned magnetic undulators needed to produce very bright X-rays, but they understood that sometimes, in order to see what something is like on the inside, you must go to great lengths to find the right method of penetrating its protective outer layers. We talked about it until my soup went cold.

I invite you to imagine reading this book as if these stories were being told at the pool or in your friend's kitchen. And if I do it correctly, by the end of it, you will have a new way of understanding your world too.

This book is thematically organized into three sections, each inspired by a different key ingredient of any physics theory or experiment: "Time," "Space," and "Interactions."

In physics, time and space are coupled into one entity: the fabric of reality that we call "spacetime." A lifetime traces out a trajectory in spacetime. Imagine a map with four dimensions—width, height, depth, and time—on which a path forms connecting all the places and times in which a person or object can exist. Practically speaking, if you are a physicist trying to calculate something, it is often convenient to "freeze" one or more of these dimensions—one direction in spacetime—and only focus on what is happening in the remainder. This allows you to take phenomena that are too complex to be analyzed with the tools we have now and reduce them to parts that you as a physicist know how to work with. Once you have that baseline understanding, you can slowly add complexity back in. In the same spirit, in this book I have separated time and space, which normally are always in relationship with each other, so that each can speak for its own complications and challenges.

In the section titled "Time," I explore memories, causality, and the way that time is both felt within us and constructed outside of us. In "Space," I deal with the challenges of being in a place and feeling that you belong to that place, of how the potential trajectories between two points can result in completely different life paths for each of us, of the possibility that many worlds may exist in parallel, and the fact that our bodies themselves envelop space.

The last section, "Interactions," deals with couplings, collisions, distant attractions, and close-range repulsions that occur when you are traveling through spacetime with other bodies. Physicists are very good at predicting what two interacting particles will do under different circumstances, but if you add just one more particle into the mix, making predictions becomes nearly impossible. When particles interact with their neighbors very strongly, the system that they comprise often exhibits strange and unique behaviors, and is resistant to many standard techniques of mathematical analysis. The same is true of systems where all particles interact with all other particles simultaneously. Accordingly, in "Interactions," I write about learning how to exist within systems of many other bodies and minds, whether that means learning how to interact warmly and authentically or letting myself enter some collisions and conflicts. These three sections complete the description of my world, and I offer them to you as a way to scaffold your own.

PART 1

TIME

CHAPTER 1

MEMORY DEVICES

"After some point, these memories, they just really suck, right? They're not useful for anything. You need to start implementing error correction, and that can take . . . a lot," said the physicist on my computer screen.[1] He waved his hands about. On my end of the Zoom call, I recognized his frustration.

"Right, right. I guess the technology we have right now just doesn't always cut it," I said in a tone that I hoped conveyed camaraderie. I had called him to discuss quantum computers and the nascent quantum Internet, which led us right into a discussion of "memories," a word that physicists use to generally refer to what someone else may call "hard drives."

The idea of building a computer that harnesses quantum effects to solve complex problems—such as understanding intricate properties of molecules that are important in industrial chemistry or medicine—dates to the 1980s. It has not only persisted but given rise to a whole industry, because these computers are expected to potentially store and process more, and more complex, information than any existing computer. This could make them powerful tools for accelerating discoveries—for example, of new drugs or better ways to build solar cells—that will have broad societal impacts. Any traditional computer, no matter how advanced, encodes information into "bits," which are made from transistors and take values of either 1 or 0. Quantum computers use "qubits," which are made from, for example, atoms or particles of light, and can have many more values.

This is because bits are binary switches with "on" and "off" states only, but qubits are more like an arrow pointing from the center of a sphere to any point on its surface—any one of a qubit's many orientations is a different quantum state that can be used to store a piece of information.

This means that quantum computers can run programs that can include more operations than their classical counterparts, where the options are only "turn 0 into 1" and vice versa. Consequently, they can run certain algorithms more quickly, or with a lot fewer hardware constraints—whereas a conventional computer may need 256 bits to complete a task, on a quantum computer only eight qubits will suffice.[2] As John Morton at the University College London told *New Scientist* in 2019, "A quantum computer is the most powerful computer that we know how to build, based on the laws of physics as we know them today."[3] Not all physicists would co-sign such a superlative statement, but in recent years I have interviewed many who have found it plausible enough to make quantum computing a huge part of their work.

But building quantum computers has so far been an incredibly difficult challenge—even technology giants like Google and IBM have not yet been able to create a device that fully lives up to the promise of "most powerful." The same is true of all other quantum equivalents of our computing infrastructure: quantum hard drives (or quantum memories) and compatible communication networks (the quantum Internet) are not yet an everyday reality for anyone but a small group of physicists.

We do have prototypes of quantum devices, including some commercial models of quantum computers that researchers across academia and industry can buy or access through the cloud. Quantum communication networks have also been tested in several big cities around the world, albeit without the full functionality of the quantum equivalent of the Internet. But, at this moment, none of them are good enough to replace their binary counterparts. The average person still does not interface with anything explicitly quantum in their daily life, and the biggest clients of quantum computing companies are still largely scientists.

Quantum computing and communication devices are also difficult and expensive to both make and operate. In 2024, I spoke to an executive at a large company who was at the time running its quantum division. The company had by then made several generations of quantum computers, and academic researchers had been using them at an increasing pace, but he told me the whole industry was still waiting for one unambiguous demonstration of a quantum computer solving some truly stunningly difficult yet useful problem. Only then would these machines go mainstream, he said.[4]

Making a quantum memory—a device that just keeps information—should be simpler than building a quantum computer, which also has to manipulate and process that information. Quantum memories and computers, however, share a fundamental problem: the quantum phenomena that make them work are fragile. This is what that researcher on that Zoom call was frustrated about. Any tiny disturbance in its environment—like someone opening a door in the lab across the hall, or the temperature jumping up a millionth of a degree—can perturb a quantum device and cause it to make errors. Imagine if your memory stick corrupted the information on it every time it was slightly rattled in your bag—you'd probably not hesitate to call it useless. Consequently, much of the job of contemporary quantum engineers lies in shielding their devices, putting them in special boxes or fridges, or trying to find loopholes in quantum physics' most basic laws to make the devices more impervious to influences from their environments. They've made some strides, but this work is nowhere near done. For instance, though you can access an IBM quantum computer online, the actual device looks nothing like your laptop. In 2022, the company invited me to see a fridge that its engineers had built while trying to decide how to best house their next generation of quantum computers, and it was so big that I could comfortably stand upright inside of it.[5]

I don't envy scientists and engineers who work on these devices. It's frustrating to only be able to keep memories intact on the condition that nothing else ever touches them.

My first memory of myself is not my own at all.

I was born in December of 1991 in a country that had been called Yugoslavia six months prior but had since become enveloped in an armed conflict that would eventually lead to our part of it being called Croatia. It was three days after Christmas, and my mom, seven months pregnant, thought her initial contractions were just indigestion from having eaten too much over the holiday. She may have indeed overindulged, yet she also gave birth to me that night, two months before her due date. But I don't know enough of the details about how she got to the hospital, or what the doctors told her, to be able to really tell the story of my birth. Instead, another incident, from a few days later, occupies the place of my origin story. It has been retold in my family so many times that it almost feels like I was there, making those memories myself.

The person who most loves to tell this story is my father, and his version goes something like this: Because I was born two months premature and during a war, he and my grandparents could not immediately visit me in the hospital.[6] I was a small shriveled-up thing with matted dark hair and tiny red hands. The doctors kept me in a special bed designed for warmth and fed me formula to help me grow. When a nurse first presented me to my father, my head fit in the palm of her hand, and my feet barely reached the crease of her elbow. Whenever my father speaks about my infancy, he always returns to this, to my diminutive size, as if he cannot believe his memory when confronted with who I am now.

When he and my grandmother went to visit me, my grandfather Branko, a man who loved routine and comfort above all, stayed at home. He spent the afternoon watching and recording a TV broadcast of *Conan the Barbarian*, the 1982 sword-and-sorcery epic starring a young Arnold Schwarzenegger. And he got so engrossed in the movie that he failed to notice when air raid warnings started going off all over the city. He missed the big yellow letters that sprawled across the top of the TV screen, urging him to run to a bomb shelter. All around him, neighbors were taking cover, but my nonno Branko could not take his eyes off Conan. He didn't even notice when it was all over, and the city collectively exhaled with relief because no one had been killed that day.

My grandfather only realized something was amiss when the movie ended and his wife wasn't home yet. Confused, he called the neonatal care unit where I was being kept. Only then, on the phone, did he learn that there had been an air raid. According to my father, everyone at the hospital had taken cover in the basement. Some nurse had taken me away from my family to wherever the babies were being sheltered. We got lucky and were reunited within a few hours.

For years, my mother claimed that she had never been afraid during those days—not about me being born early, not about the war, not even about the bombs. But recently we got into a fight that started with a failed Christmas cake and escalated toward far bigger failures in our respective lives, and she blurted out that she had been terrified all throughout the first year of my life. My father was deployed to the front soon after I was released from the hospital. Throughout 1992, as I was getting larger and stronger, he was digging trenches and tiptoeing around landmines. Everyone in my family has painful memories of the Croatian War of Independence, but the only memory they allowed to pass on to me is the one laced with humor—the story of Conan the Barbarian.

Growing up in Croatia in the 1990s and 2000s meant that I would never shake the influence of being a war baby and a post-war preteen. When I was in high school, the shadow of the war was still hovering over Croatian culture, some ten years later. Boys in my classes espoused militant nationalist sentiments as a signal that they were becoming real Croatian men. War was, and still is, an integral part of my country's collective memory. But instead of processing what it meant for my tiny body to have been enveloped in so much fear and anxiety from day one, I stuck to laughing about Conan.

I adopted this story as my narrative genesis. I told it to new friends and acquaintances, especially after moving to the United States in 2008, when being Croatian made me exotic. Because they had only experienced war as a tragedy you watch on television, however, many of my American acquaintances reacted not by laughing about Conan but by telling me how sorry they were for what my family went through. Outside Croatia, or maybe even just outside of my family, the story lost a lot of its humor. This rattled me. It rattled the fragile place in my mind where I had stored this story, like a faulty memory device being shaken. The story started to feel untrustworthy.

Now, in my thirties and several years older than my parents were when I was born, I realize that there must be an awful lot missing from it. It includes almost nothing about my mom, who must have been terrified, or my grandma, who has an even more tender heart. But for my family, accuracy may not have been the point of the story. The purpose of memories is to store information that can then be recalled later, but the purpose of stories is to make sense of that information, sometimes at the expense of precision. Most memories become stories over time, and when we tell them they are influenced by who we are, and who we want to be.

Why had my family and I been so comfortable laughing about what must have been a terrifying event? Moving away from home forced me to rethink this. It served us all to take a fearful memory and collapse it into an anecdote. We all came out looking more resilient and tough. Laughing about Conan made me sound that way from day one: the baby who'd made it through the air raid and emerged with jokes instead of trauma. When I first started living alone thousands of miles away from my family, it was useful to remember that I had once been separated from them during a war and even that ended up being just fine. Of course I would be fine this time too. It was more useful to laugh about Conan than to remember my history more correctly.

One summer when I was in high school, my family and I were driving home from a vacation when my father decided to follow the roads that passed through where he had been stationed during the war. He stopped the car a few times along the way to explain where the landmines had been and where the soldiers would gather. In dusty old Converse sneakers, I stood in places where the war had touched him. And its touch had not been light.

In the late '90s, a few years after the war officially ended, my father fell mysteriously ill with strong bouts of nausea and headaches. No doctor could name the problem, and no amount of medication could alleviate his symptoms. He would numb himself by pacing in circles for hours around our tiny living room, or by curling into a ball on the couch and tightly shutting his eyes for as long as he could. He never called it PTSD, but in hindsight it seems clear that he was suffering from it. I've never asked him whether he'd been scared when he was lying in a shallow trench, miles away from his tiny baby, or when he was separated from me in that hospital shelter during the air raid. Maybe if I had, I could have error-corrected the memories I picked up from him.

When you buy a conventional memory device, such as an external hard drive, you don't need to know what goes on inside of it to use it. Breaking it is not a matter of understanding its innards either—damage is usually caused by something dramatic, like throwing it into water or putting it in the microwave. External hard drives are typically built from semiconductors or magnetic materials, and their currents or tiny magnets are resilient to being jostled or accidentally dropped.

The quantum case is rather different. This is in part because of what the word "quantum"—usually used in contradistinction to "classical"—means in practice. "Quantum" objects earn that moniker because their energies come in "quanta," or definite parcels. A classical object, such as a ball rolling on the floor, can gain nearly any level of energy when the right force is applied to it—say, when you kick it. If you kick it too hard and then decide to kick it again so that its final energy is 10 percent lower, you can do that. However, were you to duplicate this scenario of twice kicking a ball, except now the ball was somehow a quantum object, the two levels of energy it gains would have to correspond to numbers from a predefined list—never numbers less than, in between, or greater than those listed numbers. With a quantum ball, it is possible to achieve that 10 percent reduction only if the two numbers that yield that reduction are also both on the list.

Classical objects tend to be large (that is, non-microscopic) or warm (that is, within the range of Earth's natural climate), so they comprise the familiar world that we know. To find truly quantum objects, you generally have to look to the incredibly cold, the incredibly small, or both. For example, at room temperature, a cloud of atoms—such as what comes out of a container of hair spray—does nothing unexpected. The atoms are warm, and higher temperatures correlate with higher energy and movement, so they linger in the air and jiggle in place. But if you take that cloud and cool it down to, say, -459 degrees Fahrenheit—close to absolute zero—the atoms completely change their behavior. Absolute zero is the theoretically coldest possible temperature, at which atoms stop moving completely. A place as precisely cold as absolute zero has not yet been found in our universe, though lab experiments have come within a few trillionths of a degree of it. And as atoms approach the point of that all-encompassing, unbeatable freeze, they must play by the rules of quantum physics.

This means that ultracold atoms do odd, counterintuitive things that they could never do at room temperature, like potentially being in two places at once or being something that is both a particle and a wave, but also neither.[7] And it is exactly these behaviors that make quantum memories and computers not only quantum but—in theory—more powerful than their classical counterparts. This means that, to build such devices, we really need inroads to the worlds of the cold and the small.

Luckily, physicists have known for several decades how to leverage lasers and magnets to enter the ultracold realm. One marquee year was 1995, when several research teams chilled atoms of two rare and soft metals, rubidium and cesium, to the point that they formed never-before-seen phases of matter. When you lower the temperature of liquid water, you end up with ice. When you lower the temperature of a cloud of atoms even more, you end up with something stranger than ice, such as a fluid with zero viscosity, or fluids that behave in ways with no analogue in the classical world. Those 1995 experiments were later recognized with a Nobel Prize, and they opened the door both for new explorations of the quantum world and ways to put its denizens to work.[8]

In most ultracold experiments, atoms are confined to a small glass-and-metal container without any air. Creating this vacuum inside the container is key—air is a major source of disturbance. At an atomic level, any stray particle of air can interfere with the atoms and ruin an experiment before it even gets started. The other key ingredient is lasers.

A large everyday object, like an apple or a spoon, doesn't budge when you direct a laser pointer at it, but atoms are the right size to really feel the impact of the light particles, or photons, that lasers emit. And if you hit your atoms with enough lasers, and if the lasers are calibrated to exert just the right amount of force when they collide with the atoms, all that banging around will eventually make the atoms lose lots of energy. In fact, they can lose so much energy that they reach incredibly low temperatures. Then their quantum nature reveals itself and turns them into new types of matter. Physicists can control the shape of this quantum matter by using electromagnetic forces and bursts of electromagnetic radiation, such as microwaves, to arrange the atoms into, for example, a narrow line or a flat "pancake."

Now you can start encoding information into the atoms' quantum states. There can be many of these states, and combining them creates even more options, so you now have a chance to build a powerful quantum computer or a quantum memory. The catch is that you, your lasers, your microwaves, and your magnets must be the only thing influencing what each atom is doing. Any other unplanned influence and the whole thing will malfunction. Hence the fragility of quantum memory.

I studied ultracold atoms for my doctoral degree, but I was a theorist, so I solved equations instead of building devices. Some of my peers, however, did work with those tiny rubidium-filled boxes, and I would occasionally visit their labs. I would get to peek at the confined atoms—a small box sitting on a large table covered in tiny mirrors, lenses, crystals, and lasers. Each laser beam directed at the atoms had to travel through some of those components first, to ensure that by the time it made impact it had the right properties.

My experimentalist friends always talked about the endless hours they had to spend aligning each piece of the setup. One friend told me about how his team's experiment became useless for weeks when the building across the street was under construction—even the tiniest of vibrations from the construction work disturbed the atoms too much. I became hyperaware of every edge and corner of my body, terrified that I would knock something over and set the experiment back months.

To "write" information into a quantum memory made from ultracold atoms, you again use light. You tune the properties of the lasers to encode information in their photons. When the photons hit the atoms, the collisions change the atoms' states to store the encoded information within them. As long as you manage to keep the atoms in this new state, they will hold that information. When I talk to researchers about such experiments, my job is mostly to ask about what can go wrong in this process. How many photons get lost? How many atoms change their states when you really don't want them to? How much of your information can you actually preserve, and just how easy is it to accidentally disturb your atoms?

Many existing quantum memories, even some that are considered state-of-the-art, can only hold onto information for a few seconds or less. And that is under ideal conditions—no construction across the street, no nosy theorists accidentally jostling the table.

When I was sixteen, I was unhappy. And because I was sixteen, I thought that to make myself happier I just had to go somewhere other than the 1,200-person town on Croatia's biggest island, where I had always lived. The town didn't feel big enough, nor did the island for that matter. Maybe even Croatia itself—with its population equal to about half that of New York City, and its area roughly the size of West Virginia—didn't feel big enough. So, I convinced my parents to let me apply to study abroad in the United Kingdom for "a semester or two."

The application process was not overly difficult, yet on the day of my final interview, I panicked so much that I barely remember it. My mom drove me from our coastal town to Zagreb, Croatia's capital. The roads were as clouded with fog as my mind was with anxiety. I think I was asked to write an essay before the interview, but I can't recall the essay prompt, and I can't recall what questions the interviewer asked me afterward. But I must have done well enough given what happened a few days later. I remember that part very clearly.

I came back from school on a late-night bus, got off at my stop, and saw my mom waiting there for me with our dog. I was surprised to see her—she never met me at the bus stop. Her face, framed by a halo of curly hair, was solemn. Her blue eyes were unusually hard, like a lake that hasn't yet thawed even late into the spring.

We walked toward home with a classmate of mine who had ridden the same bus. As the three of us made small talk, I tried to infer the bad news that I assumed my mother must be getting ready to share with me once we were alone. My father had been on a business trip all week—maybe something had gone wrong with his work? Or it could have been my grandma, the one who lives alone—could she have fallen and gotten hurt? Or maybe it was my younger brother, who had a tendency to get in trouble.

When we got home, the sleeper sofa in the living room was pulled out, covered in comforters and blankets. This confirmed my suspicion that a bad thing must have happened—in such moments, my family would sleep there together to comfort each other.

I braced myself. My mom began to speak. It was not about my dad or my grandma or my brother—it was about me.

I had been offered a study abroad scholarship, my mom explained, but not to the UK and not just for a semester or two. A rich American family wanted to pay for me to spend my last two years of high school in a boarding school in upstate New York. I was also provided with a list of colleges, and the family would pay for me to attend any of them that would have me. They were offering to commit hundreds of thousands of dollars to my education, a sum that I still cannot fully grasp.

"We have to let them know tomorrow—you have to decide tomorrow," my mom said shakily. I started crying because I knew that saying yes would change my life forever. She started crying because she knew that I would say yes, that she had raised me to do so. We called my dad, we strategized on how to buy some time for our reply, we cried some more, we slept on the sofa—and I did say yes.

In the years to come, that moment in which I showed so little hesitation about deciding to uproot my life became a point of conflict between my parents. In arguments with my mom, my father adopted a convenient refrain: "Now that you've let her go, she will never come back."[9] He was being unfair, but almost twenty years later I know that he was also right. I have not yet moved back to Croatia, and frankly, I don't think I ever will. In retrospect, I feel that the moment I realized that I could make such a decision was the beginning of my growth into who I am now, which is why that night is one of my most vivid memories.

Or at least I thought so, until I tried to write it down. The more I concentrated on recalling it, the more incomplete it started to feel. I can't remember what time of the year it was or what I was wearing. I can't remember whether my brother was home and if he also slept on the sofa. I must have had homework to do once all the drama died down, but I can't remember it. And I am mystified by my lack of recall about whether I called my friends or what I told them at school the next day. This memory regarding where my life as an adult effectively started has now been perturbed by all these questions.

The story of the driven, ambitious person who can leave everyone behind in pursuit of their goals is an American story, and when I came here, I really made it my own. Remembering that night as one in which I committed to leaving with no regrets or hesitations made me feel like I belonged in my new country. I could leverage it to tell myself that being here, and being who I am, was my destiny. And destiny is inevitable, even when it means loneliness and heartbreak, even when it means financial precarity, even when it means being made to feel small. Over the years, I have felt all of these things, but hey, if it was my destiny to be here, then I had no choice but to get on with it. My memory seeded a story that I could tell myself to keep going, regardless of how incomplete it may have been.

Now I can see that sixteen-year-old with more clarity. I can see their sadness, how much of it was brought on by living in a small town and trying to find meaning within a narrow-minded culture. I did well in school, but it rarely felt exciting. I had a few friends, but still often felt lonely. As most teenagers, I was searching for the person I could, or should, become, but the options that surrounded me in the form of Croatian adults only intensified my natural inclination toward melancholy and depression. Sometimes I cried, more often I sulked. Most often, I gritted my teeth and looked for myself in popular science books and heavy metal music.

My parents did their best and provided me with as much help and understanding as could fit within their worldview at the time. At a different time, or within a different culture, maybe they would have taken me to therapy or found a way to enroll me in a more challenging school or one where I would have felt less alone. My teenage angst was not exceptional, but I did very badly want to leave Croatia, which I hoped would also mean leaving behind my sadness. That sixteen-year-old, they just needed to go somewhere, somewhere potentially with more space to grow and breathe.

Sure, they had drive because they loved science, and they were resilient because that's what sixteen years of being an oddball will force you to become, but I also know that my memory of that pivotal night has been shaken up by my attempt to construct myself into a person who is self-sufficient above all. This memory had a lot of light shined on it when I argued with my family, and it was often heated when my father and I yelled at each other because we were both too proud to find a better way of saying "I miss you." This all corrupted the memory, but I could never really turn off the experiment.

Many of the researchers today who are working on building quantum memories are actually trying to build a quantum Internet. Just as the conventional Internet currently connects all our communication and computation devices, the quantum Internet of the future would connect various quantum devices to one another and allow them to exchange information by storing it in quantum states of photons. Certain quantum properties, like "entanglement"—a way of connecting two quantum particles—could be leveraged to make information in such a network essentially unhackable, which would be a boon to financial institutions like banks. (In fact, the first quantumly secured bank transfers were tested more than twenty years ago, in 2004.)[10] The quantum Internet would likely not replace the Internet that we use today, but it would usher in functionalities that are currently impossible, such as connecting many observatories across the world into effectively one ultrasharp telescope, and allowing for the exchange of highly complex information.

Many small-scale quantum communication networks already exist—in Chicago; Chattanooga; New York; Hefei, China; and a few other cities where they are being used as test beds.[11] Europe's largest seaport,

Rotterdam, Netherlands, has also "gone quantum."[12] In October 2024, a small quantum satellite connected ground stations in China and South Africa through a quantum link that spanned more than eight thousand miles, which is a record distance for sharing ultra-secure quantum-encrypted data.[13] The leap is happening, and research institutions and government agencies alike are getting on board. In 2025, I even spoke to a team of researchers who had developed an operating system, a crucial software component similar to Windows or macOS on your personal computer, for quantum communication networks.[14]

On the ground, these incipient quantum networks typically use the same optical fibers that have already been laid down for the conventional Internet.[15] Conventional networks and quantum networks both use photons to transmit information, but in the quantum networks, the photons are "quantum entangled"—that is, they share a special type of connection that always keeps their states correlated. This helps the photons carry information encoded in quantum states.

But when photons travel longer distances through the fibers—such as between two cities—many get lost along the way. In traditional networks, this is addressed by a specialized device periodically reading out the information from the photons as they travel, then retransmitting them. In the quantum case, there is no way to imitate this process without destroying entanglement. Instead, researchers solve the problem by adding quantum memories to their networks. These memories also get entangled with the photons, essentially storing information that can then be leveraged whenever a photon gets lost.

In 2020, a team of scientists in China used two quantum memories made from ultracold rubidium to show that this trick can work across about thirteen miles—a record-breaking distance.[16] A few years later, as I was reporting a story about a computer simulation of Paris as a "quantum city," fully connected by the quantum Internet, I learned that the research team behind the simulation also wanted to assess what exactly it would take to make all of Europe quantum.[17] Quantum memories were at the top of their list. Without the help of quantum memories, entangled photons will simply not be able to go the distance.

How do you make a quantum memory that is durable enough to be a part of a sprawling metropolitan quantum network? Researchers have been able to compactify some ultracold experiments so that they fit onto chips similar to computer chips, and they have also been able to create quantum memories from ingredients that are less fragile than ultracold atoms. For example, in 2015, an experiment in Australia proved that a crystal made of yttrium, silicone, and oxygen, studded with a few charged atoms of europium, could maintain a specific quantum state—and therefore a piece of information—for almost six hours without degrading.[18] Compared with existing prototypes of ultracold memory devices, which typically operate at the scale of milliseconds, this is a shockingly long time. Quantum memories that use diamonds are also improving—in a 2022 experiment, researchers managed to maintain a quantum state for more than a minute.[19]

Crystals and diamonds both harness the same process that atoms undergo when they're hit by light, but within a solid. The crystals used to construct memory devices are imperfect, or "dirty," with carefully chosen charged atoms of rare earth elements playing the role of specks of dirt. When a photon enters the crystal, this speck of dirt absorbs it, and this changes its quantum state. Researchers play a similar game with diamonds, which are just crystals made of carbon. They create a new structure at the center of a diamond by knocking out some carbon atoms, creating a vacancy, and filling that hole with nitrogen. An incoming photon then changes the quantum states of the two carbon atoms and the nitrogen atom closest to the vacancy.

Dirty crystals and defective diamonds can maintain their quantum properties at temperatures warmer than outer space, but still not so warm that they could be put in just any room. Most experiments with these types of quantum memories only work inside large refrigerators that run on liquid helium and maintain significantly lower temperatures than any household or industrial freezer. Their "quantumness" is yet again only useful on the condition that our big, warm world be kept away.

In April 2023, I spent a morning in the offices of a start-up that wants to be New York City's first quantum Internet provider.[20] Its research team was sending quantum light through a loop of fiber underneath my old Brooklyn neighborhood, and I was excited to learn that it was working better than expected. Not that many photons were getting lost, despite being in contact with the incomparable hubbub of the city above. I had previously reported on the team's efforts to build a room-temperature quantum memory that fits inside a container the size of a pizza box and also could eventually be made resilient to almost everything that goes on around it.[21] They had worked out how to get rid of any need for refrigeration but still store light's quantum states in a stable cloud of atoms. But when I asked to see this memory-in-a-box, I found out that it had reverted to being a mess of tiny components on a large table because what had fit in the box before wasn't storing information efficiently enough.

There's always a trade-off. You can never truly win when it comes to trying to do everything right with memories.

Almost everything I know about ultracold atoms, quantum states, photons, and crystals, I learned in Urbana, Illinois. Stitched to its neighboring city of Champaign by the University of Illinois campus, Urbana is the kind of flat and calm Midwestern town that I never thought I'd want to live in. But I moved there to attend graduate school, and the place brimmed with scientists that I could look up to. I developed my own routines within the town's microcosm and immersed myself in studies of the quantum world. Physics really revealed itself to me there, as did the richness of the real world when I met the man who later became my spouse. Starting in Urbana, we established a kind of togetherness that has since kept us linked even when we've been far apart. Inevitably, Urbana grew on me.

I successfully defended my PhD in March of 2020. After spending a few quiet days relaxing in my new title, my spouse and I set course for Brooklyn to spend spring break visiting with his parents there. The COVID-19 pandemic was already on our radar because classes and academic conferences were increasingly getting canceled, but it really caught up with us in Brooklyn. Circumstances forced us to move into my in-laws' basement, where we would stay throughout the summer, and only manage to move into a place of our own in late fall. All I had with me was a carry-on's worth of spring break clothing and my running shoes. The rest of my belongings sat in my empty Urbana apartment, one that I kept paying rent for until, after about five months, an expiring lease and a nervous landlord forced us to fly back to Illinois. We would stay for six days so that we could pack up my apartment, officially vacate it, and leave for good.

Masked and nervous, my spouse and I made it back on a bright hot August day.[22] I anxiously opened the door of my apartment. The mesh screen on the door was cracked and fraying, betraying the fact that temperatures had dropped to negative numbers that February. Shakily, I put the key in the lock and turned it. My living room was filled with light and everything remained as I had left it, as if frozen in time. A pair of ripped jeans was collecting dust on the bedroom floor. A bag of onions was quietly sprouting in the fridge.

We sat on the squeaky leather couch, which I always disliked but couldn't afford to replace after moving in. We tried to pretend that the situation wasn't all that different from returning from any other spring break. I found an old loaf of sourdough bread in the freezer and promised we'd have toast in the morning, just like we used to. But it all felt so foreign, like stepping into the life of an acquaintance who you'd drifted away from long, long ago.

At my spouse's old Urbana house, the stoop looked completely different. It had been repainted and no one sat there drinking bad whiskey and laughing anymore. Most of the friends we used to play trivia with had moved away, and the bar that hosted it had closed. The yoga studio where I'd spent countless hours sweating and trying to open my heart had also shut down. The farmers' market that I used to rush to before Saturday morning yoga classes was now barely a market at all—just a few scattered booths. Everything felt emptier than ever. And unfamiliar—like I had completely invented my memories of my kinship with this place. On campus, I cleared out the desk in the office that I'd shared with other graduate students in my former research group. When I looked at an old stack of homework assignments, it seemed to be mocking my inability to feel the shapes of my former scholarly life.

In the acknowledgments section of my PhD dissertation, I'd written, "More than any other place in which I have lived by myself,

Urbana-Champaign has been and felt like a home. . . . It would be absurd to thank the whole city, but with every year I have been a little more convinced that it would have been impossible for me to complete the work presented below anywhere else."[23]

On day three of the trip, I woke up to the sight of the boxes that we'd filled with books, towels, and plates, and I immediately felt ready to leave this place that was so at odds with my memories. I was angry. At the city, and at the version of me that had written those acknowledgments, the version of me that thought that piece of emotional information would withstand the test of time.

But then another day passed and then another, and just as we were getting ready to lug all the boxes to the post office, then fly back east, graduate school colleagues who were still in town started reconnecting with me, like the first few droplets of what may become a storm. A midday cup of coffee at what used to be my favorite coffee shop on campus was followed by an outdoor evening drink, and everyone suddenly had a lot to say. Mostly, we talked about what we remembered from before the pandemic and from our earliest years in Urbana. Our stories did not all exactly match and our respective memories were corrupted each in their own way, but there was enough overlap between them to keep us talking—and feel an echo of a past closeness and ease of being.

My anger lifted, and the mellowness of a small town at its best settled onto my shoulders, gently pushing them back, forcing my body to relax. Sure, everything that had happened after I left Urbana—fear, death, and the many moments of despair over my future—had made my rose-tinted memories of the place feel corrupted, but there were still conditions under which I could retrieve them. Those imperfect memories were what held my old friends and me together through the days as I prepared to leave again, the state of the pandemic still very uncertain. They gave us an excuse to laugh, and it was only in that moment I realized how much I needed that. Another gift from Urbana, Illinois, the place that had always taught me the secrets of the universe.

The trap of knowing what a good memory would look like in a box filled with esoteric materials is that you sometimes forget that people are more resilient to imperfection than the systems we dream up in the lab. People are not at all like photons.

Within a memory device, a corrupt piece of information is rarely useful. In recent years, many physicists have become incredibly focused on building devices that can self-correct their errors, but breakthroughs have only been relatively few. A lost photon is, most of the time, just lost. An accidental change in the quantum state of an atom is just that: an accident. In its aftermath, the experiment must be adjusted, then repeated. But those six days of packing up all my old life as a graduate student made me realize that I don't have to start over every time I catch an error in my memory. Being able to examine those errors can gently push me forward.

Repeatedly, I have found that the memories that I thought defined who I am, or who I must be, weren't perfect pieces of information. But there was always value in asking myself what may have influenced their origin and transmission, or what made some of their parts degrade. I can correct some errors and flush out some corrupting influences, but chances are that I will continue to cling onto imperfect memories. Because I am not a malfunctioning component in some delicate experiment, they will not render me fully faulty. The experimental run will not have to be discarded and reset. They may even afford me some laughter, especially if I extend myself the kindness of not being angry about errors.

Besides, there's still a lot of good information embedded even in imperfect memories. I've seen my grandfather's hand-recorded tape of *Conan the Barbarian*. There really was a bomb warning. And he really did just keep watching.

CHAPTER 2

INDEFINITE CAUSALITY

In December of 2019,
a bad tooth almost killed me. No—

In December of 2019,
I almost died because I failed to
take care of a bad tooth. No—

In December of 2019,
I did my best to deal with a bad tooth,
but my luck ran out in the process.

I don't remember when the pain started in one of my molars, but I remember when it got bad enough that I couldn't sleep. I was in graduate school, I was broke, and my dental insurance was so bare-bones that a visit to the dentist would have cost me more than a month's rent, so I put off getting help as long as I could. Instead, I reached for all the enabling mantras and self-deceptions around my health I had accrued over the years. Like how, because my mom had discouraged me from taking pain relievers as a teenager, my body had never built up a tolerance, so surely just a few ibuprofens would be enough to quiet the ache in my jaw, even when it got so loud that I could barely think. Like how a good three-mile run at the gym should sweat out some of the pain. Like how my teeth have always hurt a little around the stressful holidays-and-finals season and I'd always bounced back, so surely this should pass too.

But even a deep belief in willpower could not get me to sleep through the unrelenting pain that thrummed through my skull. It felt like a scream was trapped in the bone and could not escape.

Finally, I identified what seemed to be the cheapest dentist across Urbana and Champaign and braced myself for the painful bus ride to his office. I remember little from that appointment other than signing up for a payment plan that required a hard credit check and being given numerous shots that were supposed to numb my jaw but kept wearing off. The tooth was in such bad shape that the dentist could not complete a root canal, so after a few painful hours he sent me home with a script for antibiotics. Once they extinguished the bacterial mayhem that was eating my tooth from the root up, I was to return so he could complete the procedure. Delirious with numbing shots, I called a car to the wrong pharmacy, then the right one, then home, a box of pills and a pint of vegan ice cream in tow.

I had never been one to eat my feelings and in fact had wrestled with the destructive urge to count my calories throughout graduate school, but the moment felt so raw and scary that I really wanted to mimic those kids in American movies who got ice cream when they were sick. They always ended up just fine, right? I collapsed on the couch, opened the pint, called my partner of ten years, who is now my spouse, and—for the first and only time in our relationship—cried to him over the phone. He was living in a different state, attending a different school, far away from me, my bad tooth, and my pain. "I just need you to stay on the phone until I pull myself together," I said through tears. "I think it's just the drugs they gave me at the dentist," I lied. Less than a week later, I would finally see him in person from the low vantage point of my ICU bed.

Quantum physics is inhospitable to certainty and determinism. Researchers have also come to learn that it may make very little room for a straightforward sense of cause and effect.

In classical physics, the thermodynamic arrow of time safeguards causality. This means that some processes cannot be rewound without paying a price in energy, like when a cup of tea cools down after you've forgotten it on the counter. It will not spontaneously reheat all by itself; you have to spend energy to warm it again. Cause and effect runs reassuringly forward, and cannot be reversed.

Classical physics also has a universal speed limit—the speed of light, or *c* (approximately 300,000 kilometers per second). Information cannot travel faster than the speed of light, so there is a clear separation of cause and effect in time. If the sun exploded right now, the light from that explosion would take about eight minutes to reach Earth and for the devastating effects to become apparent.

However, about twelve years ago, physicists started to understand that these safeguards do not always exist in the quantum realm. "Quantum Computations Without Definite Causal Structure," read the title of one academic paper.[1] Another kept it simple with "Quantum Correlations with No Causal Order."[2] Authored by two different research teams, these papers were not declaring outright war on cause and effect, but they did signal the beginning of questioning it. The mathematical underpinnings of quantum theory—a theory that works exceedingly well in applications ranging from cell phones to particle colliders—seemed to be offering a pretty solid footing to climb onto and yell about how "A causes B" and "B causes A" need not always be mutually exclusive.

But a world where cause and effect gets muddled is hard to imagine, especially in the face of adversity or pain. In December of 2019, a bad tooth almost killed me because I waited too long to go to the dentist. It's hard to see how there could be no definitive relationship between cause and effect in this situation.

After the ice cream and the crying came the antibiotics, and they made me really nauseous. Then, there was a fever and jaw swelling so intense that I almost forgot about the pain and the nausea. Almost.

But I had made it a few days closer to winter break and to a planned trip to New York City, where my partner and I would reunite to spend time with his family. In between bouts of nausea and before the fever really kicked into high gear, I got a haircut, hoping that it would make me look less ill for the occasion. My barber, a theatrical man who told countless colorful stories of his Viking ancestry and once tried to talk me into going to hunt invasive snakes in Florida, was shocked by the sight of my swollen face. He broke character and told me as much.

I remember posting a selfie on Instagram that day. It shows me at my dining room table in a gray terry-cloth sweatshirt. One of my cheeks is so round and puffy that you'd think I had grown half a dozen new teeth instead of having had one drilled into. Over a video call, all the way from Croatia, my father told me that I looked sick.

The next day, it was cold and snowing. The frigid Illinois winter was in full swing, but I was burning up and my bottom lip had broken out into a giant sore. The swelling from my cheek was spreading into my neck, and no amount of ibuprofen seemed to be able to buy me a pain-free moment. I still couldn't sleep.

"You have been taking antibiotics for how long now?" texted my mother-in-law, a respiratory therapist. "Send me some photos of your neck," she wrote. Then, more curtly: "Go to the ER."

She saved my life. The infection that had started in my tooth was overtaking my whole body.

I have detailed knowledge of what happened next, because in January of 2020, after I recovered, I wrote down everything that I could remember, before years of telling and retelling the story smoothed out the edges of those memories and made them more palatable.[3] I know from my notes that I ended up going into emergency surgery three times, once to remove two of my teeth (the bad tooth and the one next to it) and put plastic surgical drains in my neck and jaw, once when I was intubated, and one other time, after I was already in the ICU, which even back in 2020 I had no memory of. My mother-in-law remembers it as another attempt to physically clear the infection from my body.

The first surgery was scary, but I believed that afterward I would recover quickly. I was wrong to believe that, because the next day, while I was being kept for observation, I collapsed in the shower. Then my fever returned, and I could no longer swallow, my mouth so full of spit that I felt like I was drowning. I know that I lay in my hospital bed with a dentist's suction tube in my mouth, staving off what would have been the most undignified self-suffocation. At 3 a.m., my doctors and nurses debated whether they should call an oral surgeon to operate on me again.

I think there was an EEG scan or two. I think they told me that I might need a tracheostomy, and that cutting into my neck might forever change my smile. I think I gave them my mother-in-law's phone number and told them to ask for Mimi without explaining why my own mom couldn't come. Back in Croatia, she couldn't possibly imagine how bad I'd gotten—nor could I bear to tell her. I'm still not sure who told her that, after the second surgery, I woke up in an ICU bed with a breathing tube and an opioid drip.

I remember texting my spouse, "I'm really scared." I did not have to write that detail down. That fear is permanently stored inside me, probably lodged in the soft spot in my jawbone that remained after two of my teeth were finally separated from the rest of me.

It took several days for the infection to start abating. While I was still breathing with the help of a machine, but had recovered enough to be let out of the opioid haze, a doctor explained what had happened. I had developed Ludwig's angina, he said, a rare and severe form of cellulitis that most American doctors only ever see in textbooks. To underscore the improbability of the whole thing, he added that, usually, when patients think they have Ludwig's angina, almost always it's actually something else.

The best explanation the doctor could offer me was that I had been exceedingly unlucky. I hated hearing that the cause of this nearly deadly infection might be completely unknowable.

In November 2023, I reported on a study that had found that a new kind of energy storage device called a quantum battery could be charged more efficiently if the charging process skirted the rules of causality.[4] Remarkably, this study was not fully theoretical—the conclusions of some of its complex calculations had been tested in an actual experiment. In fact, the researchers had built a device called a quantum switch that had previously been invented specifically to prove that causality can be messed with if you're good at manipulating quantum light.[5]

I solicited a comment from Giulio Chiribella, one of the quantum physicists who first drew up the mathematical blueprint for the quantum switch. In an email, he wrote, "If I think back to the early days when my collaborators and I introduced the quantum switch back in 2009, I am amazed to see how many new applications have emerged. It was clear to us that the quantum switch was opening up a new frontier and would have had many applications in quantum information, but seeing these applications popping up over the years and coming together has been a truly thrilling experience."

As a journalist, I am always asking researchers about the histories of their experiments and what exactly they have been trying to learn or do over the years. Their stories are always fascinating, but I often still find myself struggling to imagine their world. In 2009, I was a junior in high school, and the physics I was learning amounted to working out how to calculate the coefficient of friction for a ball rolling down the hill. I firmly believed that I could control my future by playing the game of cause and effect correctly. At the same time, Chiribella and his colleagues were learning that causality can be made indefinite, or unknowable. The world that he was exploring would have been mystifying to me back then, and it was still deeply confounding now.

The following description is one way to build a quantum switch, as it was first done in 2017 by a team of researchers at the University of Vienna.[6] First, you need a quantum particle whose states and behavior can be finely detected, controlled, and modified. Photons are a popular choice. They are small chunks of illumination that move incredibly quickly, pass through all sorts of objects, reflect off others, and have no mass.

You can get a single or several photons by hitting a special crystal with a fine-tuned laser, or even by just tweaking an optical fiber like those used for the Internet. Before you use the photon to test the causal structure of the world, you must determine its quantum state—a set of characteristics that will help you understand what the photon can, and might, do.

One important characteristic is polarization. If you imagine the photon not as a ball-like particle but as a tight electromagnetic wave—something that quantum mechanics happily allows for—its polarization tells you which direction in space, such as upwards or to the left, the peaks of that wave are traveling along. You can change the polarization of a photon by pushing it through a device called a polarizer—or even polarized sunglasses—and that sort of manipulation is important for the quantum switch too.

In the lab, the quantum switch looks like a table covered in a maze of mirrors, lenses, crystals, detectors, and lasers. To operate the switch, you make the photon travel through this maze to a detector. You do this with many, many photons, then tally up all the data that the detector has gathered about them. On the basis of all that data, you next perform a mathematical test to determine whether there is evidence of these photons having experienced what's called "indefinite causality": a type of causality where several cause-and-effect relationships that are normally mutually exclusive are all effectively true at once.

Several years ago, the science writer Natalie Wolchover offered an analogy for indefinite causality that I love.[7] Imagine two people, Alice and Bob, cooking dinner in the kitchen. At some point, Alice accidentally drops a plate, and this distracts Bob, causing them to burn themself on the stove and yelp in pain. But there is another scenario too, in which Bob burns themself first, and their yelp is what causes Alice to drop the plate. Within the quantum switch, each order of kitchen mishaps is true at the same time. The kitchen is a mess, but neither person can be definitively blamed.

In the 2017 quantum switch experiment, instead of the stove or the dish rack, the first object in the maze that the photon encountered was a device called a beam splitter. This device makes the photon reflect off itself *and also* allows it to pass through. It is still the same photon—tons of experimental evidence suggest that it does not break apart or split into two—but a quantum phenomenon called superposition allows it to simultaneously reflect off and pass through. This particular quantum quirk—the ability of a particle to occupy multiple states of being at the same time, or at least make it impossible for researchers to tell which state it is actually in—is crucial for what happens next.

The version of the photon that passed through the splitter now encounters a device that changes its polarization direction in a certain preset way. Next, it encounters another device that changes its polarization yet again, but in a different way. We can imagine these two successive changes being done by two people—we can again call them Alice and Bob. First Alice changes the photon, then Bob. But the earlier presence of the beam splitter ensures that there is another version of the photon in the maze, too—the one that reflected off—and this version travels a different path wherein Bob messes with it before Alice. Consequently, each version of the photon ends up with a different polarization.

After the photon interacts with Alice then Bob—or Bob then Alice, depending on which of its two simultaneous states you are tracking—it hits another beam splitter. This time, the device reunites the two versions of the photon, collapsing the superposition into a single entity. Finally, that photon reaches a detector that records its polarization. In the many times that this experiment has been performed over the years, the result has never indicated definitively who touched the photon first—Alice or Bob. In other words, researchers cannot determine the cause of the photon's final polarization—or the cause of the kitchen mess, in Wolchover's analogy.

As it traveled through the maze, the photon was simultaneously experiencing "A, then B" and "B, then A." Each of these experiences establishes a certain causal relationship. When B precedes A, the state of the photon at A depends on the effect that B had on it. You can change the fate of the photon by changing B. In the reverse order, when A precedes B, changing A can influence the proton's final state. So, superposing the two orders also superposes the two possible causalities.

A classical particle, one that behaves like a tiny ball, would reach its equivalent of the beam splitter and do only one unambiguous thing: either pass through or bounce back. It would be very simple to tell whether its final state had been caused by Alice's final touch or Bob's. With the photon and the quantum switch, though, the causality of the process is indefinite because it does not have to default to only one state at a time. If a quantum object has multiple equally likely paths to take—such as a 50 percent chance of passing through the beam splitter and a 50 percent chance of bouncing off it—it can take them all on at once, embodying a sort of "both and neither," which in itself makes an idea like causality rather blurry.

When I interviewed the lead author of that quantum battery study, he explained to me that processes in quantum mechanics are affected by measurements. It is quite normal, he said, for a quantum object to be in a superposition of many states. You have to interact with it—that is, measure something about it, as with the detector in the photon experiment—before you can start understanding it as something more definite.[8] The quantum switch shows that this is also true of causality. In some cases, no single impervious fact governs how cause is related to effect.

If I wanted to cast the story of my bad tooth as a single definitive chain of cause and effect, it would go something like this: I hesitated to go to the dentist, which caused me to almost die. I hesitated to go to the dentist because I could not afford a root canal without going into debt. I could not afford a root canal because I was a graduate student, and graduate students are poorly insured, or not insured at all, and always underpaid. Graduate students are underpaid because they work within academia, which is a system that was originally established by the wealthy for the wealthy, and in many ways has continued to operate as such. The cause of my near death is systemic and structural, and my own decision-making mattered very little. In this version of the story, I am not fully without agency, but I am set up to always end up hurt.

In the ICU, when I was finally extubated, the small breathing tube came out with more phlegm and slime than I'd thought could even fit in a person—along with a whole lot of tears. You know how babies have to cry to take their first breaths, an iconic moment in every celluloid-captured birth? It wasn't like that. I was crying simply because I couldn't stop myself. It was the day before Christmas Eve, and though I could now breathe on my own, I was too weak to do anything other than let those tears out.[9]

After the tears came what felt like a moment of clarity in which I vowed to stop doing whatever I had done to bring myself to this point. *Let's make everything different from now on*, I told myself, my throat still struggling to produce audible words. I had barely escaped a colossal danger, yet through the fog of painkillers, my mind was already telling me that I had probably caused it, so it would also be on me to prevent it from happening again. I would work less, I would sleep more, I would manage my stress more effectively, I would meditate, I would take better care of myself in every way, I would never, ever fail to prioritize my health again. Or so went the promises.

The story I told myself about my bad tooth in this moment was not the cause-and-effect chain rooted in the structural inequities of academia. Instead, it was all about my agency. I was certainly not the first or last graduate student to put off a dental appointment to save money, but almost none of them end up with Ludwig's angina. There must have been something that I uniquely added to the grim situation. Maybe I really had just made extremely bad decisions and nothing more. I constantly wondered whether I had been the sole cause of my near death.

This self-flagellation soothed me because it assumed that I had control, that in the future I could make better choices and surely prevent similar dangers. Causality had worked against me this one time, I thought, but it was perfectly reliable. I could take advantage of that.

My recovery time after the hospital lasted nearly as long as that winter break. I remember being at my brother-in-law's New Year's Eve party. I was the only sober person there because I was still taking antibiotics, and mixing them with alcohol would have made me sick. Instead of looking festive, I was looking pale and shaky in a tracksuit and a bare face. But the rest of my convalescence, which lasted two months at most, is just a sparse and fuzzy collection of uncomfortable moments.

For two months I had to do jaw-stretching exercises in which I would stack popsicle sticks between my teeth a few times a day because the surgeries had made it hard to fully open my mouth. I don't remember when I started chewing normally again, but when my partner and I celebrated my birthday on December 28, I was struggling with pita and hummus. At that New Year's Eve party, I barely finished an Oreo. I have some photos of myself from that presumably restorative and healing time, overexposed reflections of my recovering body in various mirrors. The earliest one is from late January while I was, in theory, still resting at my in-laws' home in New York. It is clear photographic evidence that in fact I was out of bed, off the couch, not actually truly healing. The stress-filled life that I had led before the bad tooth came back quickly, and I gave into it so readily that my mind barely held space to make memories of the time in between.

The intentions that I had set for myself when I'd drawn my first unassisted breath—to take it easier, to pay more attention to how I was physically feeling—all fell by the wayside once I regained some control of my body, my time, and the spaces that I could occupy. Instead of some dramatic hard reset—like shaving my head or leaving my partner or walking across a desert barefoot—I went back to work and got things done. Within a few months of being released from the hospital, I was done with my PhD: thesis written, degree defended, life pushing me forward as if nothing ever happened.

By March of 2020, though, I had been quietly rejected from enough research jobs in academic physics that I knew my career would soon be forced to take a very sharp turn. The future I had worked toward—a simple, linear progression of A (working hard during my doctoral training) leading to B (scoring a job as a researcher)—became shapeless, cloudy, uncertain. Then a global pandemic fully shattered so many desired and possible outcomes of my actions.

For a while I clung to my old ways, continuing to attempt to plot out cause and effect, but I soon grew overwhelmed by the difficulty of any predictive calculation in this new set of circumstances. Then I became something of a fatalist, believing that I would inevitably cause some terrible effect, that another crisis similar to those days in the ICU bed was always around the corner. My constant fear of making some big life error became its own kind of self-suffocation—and it was unsustainable. It made my life worse, and it made me a more difficult person to be around.

So, spending lockdown in my in-laws' basement became a masterclass in accepting uncertainty. I filled my days cooking big family meals, teaching virtual physics classes, cheering on my spouse as he completed his own degree, and trying to stay in the now, whether I was chopping vegetables or talking through linear algebra problems. Often, it felt like the arrow of time was malfunctioning—that we weren't moving forward or even backward but rather both and neither and something in between. I tried to sit in the discomfort of that. Slowly, I lost the relentless impulse to think that just one more good decision would yield me a perfect life.

Researchers have not yet built a quantum battery that can employ indefinite causal order to charge more efficiently, but the preliminary work I reported on is not the only study to show how causality-bending quantum phenomena can be useful. In 2019, a research team at the University of Science and Technology of China incorporated the quantum switch into a communication protocol called "the exchange evaluation game." Here, two players are each given a different packet of information. A third player computes a function that combines these two packets in a specific way. They win the game once the function is computed, which may take lots of communication between the three or not so much, depending on the two players' original packets. As with the Alice and Bob analogy in the quantum switch experiment, the players in this experiment were not humans but a set of mirrors and lenses, and all the information was encoded into photons. When the first two players communicated with the third through a quantum switch—which meant their messages were endowed with indefinite causality and the third player couldn't tell which order of operations had been used to transmit them—much less communication was required among the three to compute the function and win the game. In other words, using the quantum switch made the protocol more efficient. This may seem abstract, but the researchers compared their results to "striking gold" for the quantum communication networks of the future, which could run more quickly if they included protocols similar to this game.[10]

The applications of indefinite causal order may also extend beyond the gadgets we can build; they can influence our understanding of the world they exist in. Some researchers think that the phenomenon may help develop a theory of quantum gravity, which would combine quantum theory and the theory of gravity into a single theory that would explain the inner workings of everything in our universe, from subatomic particles to black holes.[11] The stakes here are high: spacetime is the fabric of our reality, and its shape makes gravity exist, but it seems that fully understanding its quantum structure is deeply intertwined with understanding the nature of cause and effect. In 2024, I spoke with two scientists who had challenged the idea of indefinite causality and offered a more conventional interpretation of the results of the quantum switch experiment. While I was reporting on their work, however, they and several other researchers conceded that even if the quantum switch can be explained away, once we figure out how to make gravity quantum, causality will again be in peril.[12]

Most of the time, I keep the story of the tooth that almost killed me to myself. In many ways, it is so severe, and so absurd in that severity, that there is no correct way for others to react to it. During the interview to obtain my permanent US residence status, an immigration officer asked my spouse about the scar on my neck, and when we told him the whole story he went so pale in the face that I briefly worried that he'd accuse us of lying.

Once, however, a family member who already knew the story asked me whether my brush with death in that ICU bed had changed me, and I realized that I had never thought about that part—about having escaped death. This invited a third version of my bad tooth story.

In this version, I made a great decision when I told my mother-in-law about my illness, and another great decision when I went to the ER on her advice instead of second-guessing her or turning to WebMD. In this version, my recovery is swift because I spent so much of my twenties running, practicing yoga, eating vegetables, and otherwise making my body stronger and more resilient. I manage to regain my health because of my earlier choices to surround myself with people who care for me enough to engage with my doctors and sit by my bedside for days on end. The result here is me surviving, recovering, getting stuff done—being well rather than dead. And I am, just like in the second version of the story, the cause of this effect.

In reality, I probably need to look to quantum theory for inspiration and assume that all causal chains of events are true while, at the same time, none of them can be singled out as the unambiguous truth. The three stories of my tooth coexist within me now. Sometimes they jostle inside my mind in a way that feels dangerously close to some sort of rupture. I get indecisive—torn between feeling angry with my past self, thankful for their decision-making, or convinced that they were simply exploited and unlucky. In those moments, I'm grateful for all those photons that managed to keep themselves together while experiencing multiple causalities at once. My body and I are still learning how to do the same.

CHAPTER 3

PINK NOISE AND THE ARROW OF TIME

In classical physics, the first person to introduce the idea of time having a preferred direction was the British astronomer Sir Arthur Eddington. In his 1928 book *The Nature of the Physical World*, he asked the reader to imagine an arrow pointing in the direction of the passage of time—the "arrow of time."[1]

The arrow of time is peculiar in that it doesn't operate at the level of our world's smallest building blocks. Mathematically, the microscopic laws of physics are fully reversible. These are also the laws of quantum physics. If you're a single atom trying to follow some physical edict, such as becoming more energetic, and then someone plays your life's trajectory in reverse, you wouldn't notice anything amiss. But Eddington noticed that this reversibility breaks down when you consider more than just a few atoms. If you have enough atoms to be able to talk about their entropy—which enumerates how many different, messy configurations they can assemble themselves into—then you can say that the arrow of time points parallel to the direction of increasing entropy. Are your atoms getting farther and farther away from a single best-organized arrangement? Congratulations—you are moving forward in time.

Eddington's reasoning was based on more than entropy—he also considered our own perception of the passage of time. The arrow of time is "vividly recognized by consciousness," he wrote, adding that "it is equally insisted on by our reasoning faculty, which tells us that a reversal of the arrow would render the external world nonsensical."[2]

In 2020, a magazine editor raises this point about time's forward orientation after I suggest writing about researchers who were trying to find evidence of an arrow of time in classical music. The editor is skeptical that this research constitutes an interesting story. "Someone formally proving that music is irreversible in time is not surprising," he writes to me in an email. "If it were reversible, it would indeed sound like nonsense," he continues, unwittingly echoing Eddington. I convince him that there's still an interesting story here in the meeting of music and statistical physics, and I refrain from saying anything about the kind of nonsensically noisy music—such as black metal and death metal—that I like. When I get the researchers on the phone, I learn that, in fact, it took them a whole lot of mathematical intervention to prove something that I almost conceded to calling obvious.[3]

"The Jack," the third song on AC/DC's 1976 record *High Voltage*, is about gonorrhea, or "the jack," in Australian slang.[4] In particular, it is about a man getting gonorrhea from a promiscuous woman. The lyrics use playing cards as a thinly veiled metaphor. Vocalist Bon Scott, whose voice always sounds dirty and mean, snickers about how "she'd been dealt with before" and how she "said she'd never had a full house," ultimately framing the titular Jack as the card with which this woman would bring him down.[5] At the beginning of the song, Scott sings in a relatively low register, almost like he is dictating a letter to *Penthouse* over a classic bass line, but he becomes higher pitched and more agitated as the song nears its conclusion. "The Jack" ends with a shriek and a barrage of drum noise.

I was introduced to "The Jack" by my father, as was the case for most noisy things from my childhood—he was always a big source of household noise. Partly, this was because he yelled when his favorite soccer team was losing, and also when they were winning. Partly it was because he was the keeper, curator, and purveyor of noise-filled tapes and CDs. "The Jack" was featured on one tape that we often listened to in the car. Because of those car rides, *High Voltage* is still my favorite AC/DC record. It is raw, unfiltered, and minimally produced. It sounds like it was made for dingy bars full of men drunk on beer. Most of the band's later records feel a lot more optimized for arena shows, which AC/DC still play. During those childhood car rides, *High Voltage* sounded exhilarating.

We were always rushing somewhere, always getting in the car, and always putting on loud music, even for just the fifteen-minute ride to my grandparents' house. My dad and I would bang our heads to "Hells Bells" and "T. N. T." on the way to a family function or while running errands. I didn't know about dingy bars and beer, but I surely knew that we were having fun. I still cherish these times with my dad, and I loved hearing him tell stories about these rock stars that he had followed for years. I sang along to AC/DC long before I understood a single word of English. In between songs, my dad would tell me about seeing AC/DC in concert when he was younger, about how alcoholism had led Bon Scott to an untimely demise, and how Angus Young, the band's guitarist, often wore a schoolboy uniform on stage. He was also known for mooning the audience. Seeing his ass in a concert video was probably the first time I'd encountered such crass nudity from an adult and not just immature boys at the beach.

I took my first English class in the fourth grade and my mom, an English teacher, never missed a chance to expose me to the language at home, so I was not yet even a teenager when I started trying to decipher the lyrics of my favorite songs. When I gleaned that "The Jack" had a storyline in addition to the riffs, my dad explained that it was a song about gambling. Maybe he was trying to avoid talking to me about sex and misogyny. Maybe his few years of shaky high school English classes several decades prior hadn't given him enough to break through the language barrier, so the song's meaning was obscured from him, too. Maybe he didn't want to dispel the rock 'n' roll magic that we so gleefully shared. In any case, once the Internet made it to Croatia around the time I was in the sixth or seventh grade, I looked up the lyrics on our dial-up connection and put it all together.

I wasn't shocked. AC/DC songs are mostly about either playing rock 'n' roll or fucking. Some of the women in their lyrics are given a bit of deference—as in "Can I Sit Next to You Girl" and "Little Lover"—but most are painted as complete vixens, as in "You Shook Me All Night Long" and "Whole Lotta Rosie." There really isn't a Madonna/whore binary in AC/DC's oeuvre, because there are no Madonnas to be found. I think my dad understood that much.

Though AC/DC's music is noisy and loud, each song is laid out in a pattern that is predictable and easy to follow. Verses of simple storytelling alternate with powerful choruses that eventually repeat for added drama, building the song's crescendo. In an AC/DC song, a guitar solo never catches you off guard; you can always guess when it's coming, and when it hits, you feel it in the pit of your stomach and the palpitations of your heart.

Physicists and mathematicians can quantify this predictability by examining the mathematical structure of the music. Seen as a collection of aural signals that follow each other in time, a musical composition can be rated on how random it is—or, in physics parlance, how much entropy it has. A perfectly random composition—such as a collection of tones that does not follow any pattern, with any tone being an equally likely successor to any other—would be the least predictable and would have the greatest possible amount of entropy. A perfectly predictable composition, such as a single note repeating over and over, would be the least random and only have a minuscule amount of entropy. In the 1920s and '30s, some modernist composers experimented by serializing musical elements such as pitches and rhythms, which could produce low-entropy music. In contrast, others experimented by giving each tone in their composition equal treatment rather than composing in a key—Arnold Schoenberg's twelve-tone technique is a well-known example—which could result in chaotic-sounding music with high entropy.

Music theorists have used physics concepts like entropy to capture the essence of these styles, and how their discontinuous and pointillistic qualities set them apart from more traditional composers. "Although entropy results are numerical and objective, they are in some sense metaphors for a musical experience," writes Sarah Elizabeth Culpepper in her analysis of the works of the composer Anton Webern.[6] Webern was especially interested in the mathematical structure of his compositions and endowed some of them with a certain kind of mirror symmetry, as in Symphony, Op. 21, for instance, where the second half of the twelve-note series on which the composition is built is a transposition of the first half. The two halves meet in the middle in a way analogous to a perfectly blue sky touching the perfectly blue water of a lake: the sky is mirrored by the water below the horizon, the water is mirrored by the sky above the horizon, and the scene would be the same if the two exchanged places. Accordingly, scholars have described Webern's compositions as lacking "any 'law of gravity' to keep things right-side-up."[7] Culpepper argues that predictable, low-entropy musical works may be easier to process for listeners and therefore shorten their subjective experience of the length of the piece. Here, the idea of entropy sits at the intersection of physical time, as captured by the arrow of time, and an individual's psychological time.

In the 1960s, some minimalist musicians experimented with extremely repetitive and featureless music—again a low-entropy endeavor—in order to invoke the opposite effect: one that was hypnotic and nearly psychedelic. Writing about composers such as La Monte Young, Terry Riley, Philip Glass, and Steve Reich, music journalist Harry Sword notes that their work "hinged on stasis" and relied heavily on repetition.[8] In Young's band the Theatre of Eternal Music, for instance, many of the members simply held a single pitch on their instrument for as long as possible, creating droning sounds "unadorned by any rhythmic accompaniment."[9] The result was mesmerizing—in Sword's words, "an attempt to tune into the vast and unknowable cosmic infinitude"—and very, very noisy.[10]

Physicists often use the word "noise" to convey a sense of high randomness—as with white noise, for example, which has no hierarchy among its sounds and holds all frequencies at the same intensity. In white noise, there is no inherent order or structure from which a listener can infer some linear narrative. Most composed music, including many classical works, can also be classified as a type of noise called pink noise. Unlike white noise, pink noise does admit a hierarchy among its sounds—the amount of energy the sound carries is the same for each octave interval instead of being distributed fully evenly. Each type of noise is named after a color as it would appear to us if the noise was made from waves of light instead of waves of sound.

I learned this while working on that story about music and the arrow of time, during an interview with a researcher whose favorite band was Rush.[11] Rush is nerdier and more mathematical than AC/DC, but less random and dissonant than the avant-garde death metal that I started to like after I grew out of my father's favorite bands. Although this researcher and I didn't have exactly the same tastes, I still felt a kinship with him as he explained to me how the arrow of time is hidden within the works of great Western composers like J. S. Bach and Niccolò Paganini. Both, he told me, were purveyors of pink noise.[12]

This was one of my first freelance science writing assignments, and after talking about the mathematics of this for almost an hour, I still worried that I didn't understand it all well enough to write about it. But the main takeaways from the calculations resonated with me: essentially, the structure of composed music creates a kind of noise that is just unpredictable enough to keep the listeners interested, but not so unpredictable as to make them uncomfortable. "The composer of a melody must make the entropy of his music low enough to give it an apparent pattern and at the same time high enough so that it has sufficient complexity to be interesting," wrote Richard Pinkerton in *Scientific American* in 1956. After analyzing a few dozen nursery rhymes that were popular at the time, he found he could emulate them by using a composition process that included a mix of rules and a random process of drawing cards for each consecutive note. He seemed to show that these simple songs contained a mix of predictability and surprise that could be captured in a nearly algorithmic way.[13] But there is more to composed music than a mathematical balance. The structure of music also serves to build a narrative that moves forward in time just distinctly enough to be irreversible—many composed pieces would not make sense if you tried to run them backward. It is a structure that forces you to recognize a story with a beginning and an end.

When I was in high school there was a common joke about death metal among me and my friends, all of whom were self-proclaimed metalheads.

You'd name an extreme band, usually Cannibal Corpse, which even then was controversial for its violent imagery and graphic lyrics, and claim that playing its songs backward wouldn't make a difference—that's how evenly and relentlessly grating their records were. We would all laugh, not so much at the musicians, but really at ourselves, with our all-black outfits, big hair, and penchant for getting into mosh pits at the shows of any loud band that came to Croatia. Cannibal Corpse, however, wasn't my favorite, and I preferred the "melodic" subgenre of death metal. There were the early records by In Flames, Dark Tranquility, and Opeth that mixed fast drumming, strong riffs, and growled vocals, which often resembled guttural screams more than discernible words. Over the years these bands softened and moved closer to the mainstream, becoming more singsong-y and easy to follow—and I went the other way.

Cult of Luna's album *Vertikal* mixes death metal and more electronically inflected noise into long, droning songs that make your stomach curdle with premonitions of a dystopian, machine-laden future, and it was an incredibly formative record for me when I first heard it in college. When I saw the band perform it live about two years later, it made my whole body reverberate. The noise was so thick that I could have shrouded myself in its intricate sonic weave. In 2013, I was similarly enamored with Deafheaven's *Sunbather*, whose vocals are almost as screechy and diffuse as white noise and the drumbeat that undergirds them is not a gallop but the world's longest sprint. But the album is also romantic and tender; there is softness hidden in the layers of sonic abrasion. When I bought a physical copy, it was pressed onto pink and orange vinyl, almost foreshadowing my later discovery of music being pink noise.

Early on in our relationship, in an act of romance of his own, my spouse took me to see the band Thou perform in New York, and the acidic edge of their own fast and grating style of metal, a sort of toxic sludge for your ears, also got me hooked. I was already in graduate school and one of my favorite ways to procrastinate was to make playlists with descriptive titles like "Desolated," "Grating," and "Technical/Brutal/Death." Each offered a different flavor of music dense with tones and rhythms, always built on top of a drum sequence busy enough to feel like it's edging on random, always finished off by yells, croaks, and every other noise that could not be passed off as singing in any context other than metal music. Bands such as Ulcerate, Phobocosm, Decapitated, Bloodbath, and Obituary all delivered on their names, serving up music that was not without narrative, but for whom that old joke still kind of held true—had you missed a part of a song or mistakenly played it at the wrong speed or even backward, the result could have still been concordant with the rest of the playlist.

And then there was Sleep and Boris and my ongoing love affair with doom and drone metal, two genres that taught me to let music wash over me instead of always expecting it to come in headbanging-friendly bursts. Listening to Sleep's *Dopesmoker* feels like swimming through molasses. The band had never been shy about how much of their music is related to drug use, but I found it intoxicating all by itself—it made me feel high long before I ever considered achieving that feeling with chemicals. *Feedbacker* by Boris, an even slower, even noisier album, worked on me similarly, holding me transfixed, activating some part of my consciousness that did not need a neat narrative, nor a clear story told linearly in time, to feel content and pleased. While I was finishing this chapter, my spouse and I went to see The Body, an experimental metal duo that I can only accurately describe as a drummer and a guy who screams while playing guitar. As I watched a wave of bobbing heads and swaying bodies spread across the crowd in the back room of a Brooklyn bar, I wondered what a mathematical analysis of the sounds that caused that wave would show. Could this still be the same category of noise as Paganini or Bach? The thought evaporated from my mind as the barrage of sound suspended me in an extended moment where I could feel a lot more than I could think. Whichever part of my teenage self is still left within me was glowing. Until the band's set concluded, the physicist in me, on the other hand, had uncharacteristically little to say.

Yet, when I think of the whole arc of my life, physics always seems to have something to say about my tendency to mark different points along my own journey through time with associated pieces of music. I think of large swaths of my life as divided not into years, but intense periods of listening.

Along my path through time, you could lay down one marker on the years spent listening to "The Jack" with my dad in the car, banging my head to the riffs while oblivious to the lyrics. A second marker could be laid many years later, when my father and I were again listening to "The Jack," this time while playing cards of our own with my husband and my brother. It was my first visit home in a long while, as my graduate schoolwork, immigration bureaucracy, and the COVID-19 pandemic had kept me in the United States for several years, and it was my first visit home as a doctor and a married person. Throughout the trip, everyone kept congratulating me on becoming a "Mrs." but not on becoming a "Dr."

It was summer, and we were spending most of our time in my family's cabin in the woods where the sun was less punishing, everyone was more relaxed, and card games were nearly mandatory. My father's family built the cabin long before I was born, and I was just a toddler the first time I visited it. It became a staple of my childhood, and the older I've gotten, the more fond of the cabin I have become. I cherished the card games so much that I later taught them to my American partner. But now, as "The Jack" unfurled its predictable and familiar structure, I felt a rift open between me and the men at the table. Bon Scott didn't even have to shriek to push my growing discomfort to a climax. My father beat him to it by using a crass Croatian idiom to refer to the unfavorable cards in his hand. He called them "whores." Despite now being someone's proper "Mrs.," I felt even smaller than I had been when I'd first heard "The Jack." Whatever flavor of sexism I may have been protected from back then was now inescapable.

You cannot play "The Jack" backward because flipping its arrow of time would lead to nonsense. The personal narrative that formed between these two moments for me, marked by the same song, also felt thoroughly irreversible.

"Innuendo" is the title track of a 1991 Queen record, the last album the band recorded before its lead singer, Freddie Mercury, passed away about ten months later from complications of AIDS. The song opens somewhat ominously, with a heavy march-like beat and a reverb that evokes Queen's earlier dalliances with science fiction. Then the song swells into something between optimism and a soft battle cry, all delivered with Mercury's signature theatricality.

While recording *Innuendo*, he almost certainly knew that he was dying. There is no trace of weakness or fear in his voice, but the lyrics are too potent to be coincidental. "Innuendo" is filled with urgency and resolve as Mercury repeats, "Oh, oh, we'll keep on tryin' / Till the end of time."[14] The song also weaves a narrative of revealing yourself in your truest form, of releasing any fears of being known. One interpretation, albeit somewhat crude, is that in recording this album, Mercury was trying, and trying to be true to himself, till the end of his own time.

My teens and early twenties were all about metal music, but before that I was absolutely obsessed with Queen. I listened so relentlessly to the "best of" mixtape that my dad had made that my mom actually went out and bought me some pop music CDs to listen to instead. Her effort failed. I kept blasting "Crazy Little Thing Called Love" and "Spread Your Wings" over and over. Once, I woke up with the sun on a weekend morning to watch a rerun of a documentary about Freddie Mercury on Croatian TV. My parents, who always struggled to get me out of bed for family gatherings or church, were shocked. Afterward, I repeated what I'd learned about Freddie to anyone who would listen. I sang along to all of Queen's slow songs, hoping that some of his cool and appeal would transfer onto me.

At the time, my favorite Queen records were *News of the World* and *The Game*—I had not yet heard *Innuendo*, and only learned about it from that documentary. I remember an interview with a band member, who looked straight into the camera and somberly recalled how, during the recording of *Innuendo*, Mercury would take a shot of liquor, briefly snap out of his illness, and sing his heart out in the studio. His bandmates were amazed. I'd learned early in my fandom that, like many queer and gay men of his time, he'd died from complications of AIDS—the media liked to mention this fact to add drama and scandal to his story—but I'd never paid much attention to it. When my dad said that Mercury was one of the few men in history who got a pass for being queer, I felt my stomach flip but did not really understand why. It would only be at a different stage of my own personal arrow of time that I would return to Queen and realize that what had drawn me to Freddie was likely a kind of queer kinship. When I'd been growing up as a young, tomboyish girl in a conservative small town, the possibility of seeing myself reflected in him had been overpowered by popular culture's insistence on viewing Mercury's sexuality only through the lens of shame and death.

Today, I can read as much queer subtext into Queen's music as I like or need to, and that reading would probably be rarely at odds with the truth. But it is also a more general feature of composed music that listeners grasp for intentions and meanings regardless of whether those align with the author's intent. A version of that feature is also at play with the researchers who study music from mathematical and physics perspectives. Music's arrow of time seems especially conducive to inviting speculation on what it is that composers actually do.

This came up in my conversation with that Rush-loving physicist. His team had analyzed about eight thousand pieces of Western classical music. Their dataset included many different composers' works, and for each, the researchers calculated and compared several statistical properties. Ranking compositions by just how irreversible they are uncovered differences between the flashy Italian violinist Niccolò Paganini, the Romantic opera composer Georges Bizet, and the big orchestra lover Gustav Mahler.

In their paper, the team included a simple chart ranking the three composers on how time-irreversible their style was.[15] Bizet beat Mahler, who, in turn, had created music that was more irreversible than Paganini's. I marveled at this. Mathematical concepts like entropy had been used to classify genre—for instance, to distinguish the more repetitive beats of electronic music from the more varied riffs in metal and punk—and that made sense to me.[16] But when it came to these composers, without the aid of mathematics, I would have not gotten past the very simplistic observation that none of their works would make sense when played backward. Could a person really discern musical gradations in time irreversibility? Do composers know just how much of the arrow of time they're building into their works?

I was not the first person to be struck by this. The composer and music theorist Iannis Xenakis, in his 1963 book *Formalized Music*, directly posited that "much like a god, a composer may . . . invert Eddington's 'arrow of time.'"[17] He used matrices and differential equations to analyze music and create pieces that were stochastic and irreversible by design. Scientists, however, tend to be very careful about interpreting their calculations of music's irreversibility. While reporting on the arrow of time, I kept asking physicists to speculate about composition as a process of manipulating information or constructing a system, but all the interviewees politely did the safe and scientifically correct thing and pointed me to their plans for future research.

Several years later, when I was writing about a mathematical analysis of the oeuvre of Johann Sebastian Bach, a researcher helped me better understand how a mathematical analysis can be rigorously connected to how people actually engage with music. First, you convert the music to math, then you convert the mechanisms of human perception to a computer model, then you feed the math into the model, then you try to validate the results by having real people listen to the real music while getting a brain scan, then you iterate until all the models are perfectly in line with real life.[18] It was a lot. It also still didn't come close to fully unraveling the process of composition. Modeling how people perceive music was already difficult enough.

"Will this really allow us to understand every detail of how humans and music interact?" I asked. "I have to say yes, because that's what I'm betting on," the researcher retorted with a smirk before quickly hedging. "But there is so much more work to be done," she said.[19]

Regardless, when I hear a song that instantly takes me into the past, or that glues together two disparate points in time, I think about composers as time-arrow-makers. In my mind, they are superpowered people capable of grabbing disparate, disordered pieces of a story and arranging them into a structure that surprises and soothes in perfect proportion.

In my early twenties, as I embraced heavy metal, death metal, and thrash, I abandoned Queen as a soft, frivolous interest of my childhood. I only returned to them as a somewhat disgruntled adult, after realizing that harder and heavier are not synonymous with more virtuous, and that the metal community was—at least in the mid-2010s—not exactly built for people other than straight men. While working on calculations for my PhD late into the night, I'd pull up Queen's 1986 *Live at Wembley Stadium* on YouTube, hoping that its energy would keep me awake. Most of the time, however, my equations would go unsolved as I got emotional watching Freddie prance around the stage, awash in the crowd's adoration.

Freddie in the white track pants with the red stripes. Freddie in the short yellow jacket. Freddie in the white singlet and sweating. Throughout that show, he is an icon from the very first second to the last. As I grew older, I could recognize that he was also not just any icon, but specifically a queer one. When I moved to New York, seemingly every butch queer woman and transmasculine person I met admitted to having once worn that white singlet outfit as a Halloween costume. I now have friends who call me "cutie" in the same cadence with which Freddie greets the Wembley audience as he says, "Hello again, my beauties."[20]

I have long understood that musicians that I adored as a child, such as AC/DC, never wrote their music with a fan like me in mind, and that I would have probably felt unsafe had I found myself in a bar full of their acolytes. Queen felt different. Now I understand how, as a child, I grew so obsessed with them and specifically with Freddie, who never became a romantic crush. I wasn't drawn to him because I was in love with him; it's just that the seeds of what drew me to his energy—a singular queer masculinity and power—were already within me.

One dramatic plot point in the Freddie Mercury documentary from my childhood took place in the 1980s, just a few years after Queen's staggering breakout success with the hit song "Bohemian Rhapsody," when Freddie cuts his long hair short and ditches the flamboyant and feminine trappings of glam rock. As a kid, I had sung along to Queen records and imagined being famous. If I were to imagine becoming famous now, I would certainly make space for the dramatic documentation of the moment when I too turned more butch.

But "Innuendo" also gets its own moment in my story, and its own performance. In this case, however, the year is 2021 and the performer is my mom. We are attending her pole dancing class in Croatia, and I am crying.[21]

She was a regular in the class; I was just visiting and trying to keep up. At the start of the class, we did a mix of yoga and calisthenics, neither of which was exactly foreign to me but now I was doing them outside my usual context. Unlike at my New York gym, here, I was surrounded mostly by women in their forties and fifties, mostly in outfits that were skimpier than my own regular workout clothes. The class evolved from there, first with a moment of spiritual affirmation and talk about the divine feminine, then the practice of some very technical moves—one of which was new to the class and included hanging off the poles—and, ultimately, actual dancing. All the women seemed to move through the class with ease and comfort, immersing themselves in this mix of physical training, new-age sisterhood, and tapping into their sensuality. All regulars, they joyously stretched, crunched, and rolled, demonstrating a confidence that made me jealous.

The physical strain of the class didn't bother me, and the pole dancing move that was being taught that evening wasn't beyond what my body—used to strenuous exercise—could handle. I could do the parts of the lesson that required being strong—flexing a leg, broadening the shoulders, tightening the core, gripping the pole. What I could not do was be confident and elegant, the part that would have made it dancing and not just working out. Women spread around several poles in the studio and alternately practiced the new move and cheered each other on. My mom spun and swung on her pole easily and her cheers were natural and genuine. She fit into the space just right and was enjoying every second of being there. Excitement and joy emanated from her brightly.

When it was time to dance, the class moved to another room where there was only one pole. From a small bag that the instructor referred to as "cosmic," women drew slips of paper with the titles of songs. She called out a specific student to dance, one by one, while the rest of us sat on the floor and watched. That night, the cosmos wanted my mom to dance to "Innuendo."

Before starting the music, the instructor invited each dancer to consult her inner "yoni," or source of female power. The divine feminine had already been a prominent character in the arc of the class, and the instructor kept implying that this energy was bringing us all together.

With enthusiastic earnestness, and at least some disregard for the way in which she was borrowing spiritual practices from other cultures, she sold the whole thing as honest female empowerment, tinged with a sense of the mystical. The women ate it up. They felt powerful. I only felt my body contract—it was like every one of my muscles wanted to curl up and shrink out of sight. I was grateful that, as a newcomer, I hadn't been asked to dance, as the discomfort that was growing within me would have made doing so absolutely impossible. The instructor dimmed the lights.

When it was my mom's turn, "Innuendo" rang out and, bolstered by Freddie's voice, she worked herself into a frenzy. She soared with the song, moving with its beats as if they had been made just to serve her body. "Release your mask," Freddie implored, but she had never even brought one. I huddled in the corner next to a woman sporting several different layers of fishnets, my face growing wet.

I was crying because I was moved by my mother's effervescence and might. And I was crying because of the sense of difference between us; another gendered rift. The grief that this invited felt so tangible that I had to push it out of my body, transmuted into water and salt. There was no divine feminine in me, no female energy, no inner goddess. Wherever it was inside of me that my yoni should have been, all I could find there was Freddie in the track pants, strutting across the Wembley stage.

Later, while we were driving home, my mom told me that she had never felt energy or a sense of power inside of herself when she was younger, but that feeling it now had changed her completely. Her personal arrow of time had taken her from a state of obscured female energy to a state where that energy was everywhere for her, where she was basking in it. I can attest to it: now she cannot enter a room without ushering in awe and glamour as well, and there is power in her colorful dresses and big hair. And she is filled with expressive passion—the protocols for pretending you don't care are simply not coded into her body. At the edge of sixty, she looks remarkably young, like we could be siblings. Yet, I am certain that I am not made of the same stuff as her.

In the car after class, presuming that she knew how I felt, she told me how she'd been self-conscious of her body in the past and how that may have held her back. Having not yet found the confidence to tell her that I felt nonbinary, I understood why she was saying this. I understood that she wanted me to realize that I could find a way to free myself and find my own flavor of deeply embodied female empowerment. Maybe I would have made myself believe that too, had I not seen her dance.

So, another pair of markers along my arrow of time: loving Queen when I was too young to understand the word "queer," and loving the band again in that dance studio, when I very much did. Here, too, the arc of my life in between the two markers felt firmly irreversible.

Eddington's discussion of the arrow of time is now almost a hundred years old, but the question that it raises—of how this arrow emerges from laws of physics that are indifferent to time's direction—remains open. What is especially troublesome is that these are laws that apply to the smallest building blocks of our world, or the quantum particles that everything is made from. In 2014, an international team of theoretical physicists suggested that the quantum property of entanglement may be a big part of the answer.[22]

They focused on the fact that the arrow of time points in the direction of increasing entropy, and also toward a state of equilibrium. A warm cup of coffee cools down to equilibrate with its surroundings, but a cup at room temperature never spontaneously warms up because that would mean leaving equilibrium. This preferential movement toward equilibrium breaks the symmetry of the two possible directions of time, distinguishing forward from backward. Across several mathematical investigations, focusing on quantum objects, the team argued that what makes an object reach equilibrium is that it becomes increasingly quantum entangled with its environment.

Therefore, the arrow of time points in the direction of an increasing number of entangled objects. If the atoms comprising a coffee cup are entangled with just a few air molecules in your kitchen, you must be observing it at an earlier time. If they are entangled with every atom in the room, you are observing it at a later time. A similar argument may explain how humans form the idea of "now." According to this framework, we can only remember something after we become entangled with it, or with the particles of light and sound that communicate its presence to us. Physicist Seth Lloyd once summarized this by saying, "The present can be defined by the process of becoming correlated with our surroundings."[23]

As internal as our sense of time may seem, a more esoteric idea looms large here. It takes just a little propensity for romance to read this theory as the universe imbuing each of us with a sense of time by entangling us into a constantly expanding network of connections—by smearing out the edges of our independent existence, making us inextricably connected with everything else.

This take on the arrow of time, though supported by mathematical arguments, never gained consensus among physicists. It simply joined several other theories concerning the passage of time, including those that posit that we live in a "block universe" in which everything is actually unchanging and in which the arrow of time is either a mental construct or a quantum mechanical illusion. In his 2023 book, *Molecular Storms: The Physics of Stars, Cells and the Origin of Life*, Liam Graham lays it out simply: "Observers—such as you or me—are collections of molecules in a far from equilibrium state. For such a state to exist the universe must be out of equilibrium. Therefore, the second law means it is tending towards equilibrium and so there must be a thermodynamic arrow of time."[24] In other words, he centers the fact that we exist as part of the reason why we also see time moving forward. When it comes to the arrow of time, this is what the block universe theory and the quantum correlation theory share: both point the finger at the inextricable connection between people and time, which we seem to be stuck having to experience.

I think this is why I keep coming back to music as a container, or a code, for a miniature version of the cosmic arrow of time. In a mathematical sense, the arrow of time exists in music because music is a structure made from sounds and the transitions between them. As a universal reality, this doesn't have to do with me. But on a more personal level, when that arrow takes me through a familiar song, I have to confront my own passage through time and who I was when I heard that song for the first, second, third, and every other time. I can return to one of those past versions of myself, and how that person was connected to the rest of the world and their own loved ones, while also acknowledging that though the song can be rewound, my own motion toward some distant equilibrium cannot. Within a few loud minutes, time and I grapple in the most interdependent of ways, feeling out each other's soft, moldable edges. Neither of us ever really wins—we only grow more intertwined. And the process gifts me just a little more clarity on who I may be in the place where the arrow of time pushes me next.

My partner and I have always bonded over music, though his taste ranges far more widely than mine. I recently asked him to recommend some songs that deal directly with time. He rattled off about a dozen. "Isn't the passage of time what most music is actually about?" he asked, looking at me with soft eyes. I bit my tongue. There was no need to bring mathematics into the moment.

PART 2

SPACE

CHAPTER 4

THE MANY-WORLDS INTERPRETATION

You are here to do science. People tell you this all the time. Their words may be stating something else, such as they are craving pizza, or just bought a sewing machine, but behind them you always hear a whisper reminding you that none of those things are what you are here for. What you're here for is to do science.

In the mornings, you dangle your feet off your unreasonably high bed, above the gray floor that never looks clean, and get mentally ready to face the day and do science.

Twice a week your breakfast is a roasted sweet potato with chickpeas and tahini, and the other five days you have a green smoothie. You are really committed to meal prep, probably because it makes you feel grounded and in control. You eat alone, sitting at a small circular table, a small cup of Croatian-style coffee never not by your side. The cup is made from a yellowing plastic material with fading palm tree illustrations. Phone in hand, you study social media posts of fitness and wellness influencers fueling up for the day. You're doing it too: getting your protein, fiber, and healthy fats. You're not sure what the influencers do all day, but you know that you do science.

The face that looks back at you from your three-part bathroom mirror often betrays that you are tired. It also begs some questions. Is this what a scientist looks like? What does a scientist look like? Every morning, you decide that scientists wear black eyeliner and heavy brows. Most days, you wear your uniform: a yellow cardigan and a red bandanna. You used to be more extravagant, all puffy skirts and short shorts, but then a student wrote that you "dressed inappropriately" in an evaluation, so you toned your clothes down. You sort of hate that you did this, but you've been around for a while, and you are tired, so you have to conserve your energy and pick your battles. Otherwise, how are you ever going to manage to do science?

You always rush out of the door, as rushing has become your natural state. Most days you rush to your office, where you are usually alone even though two other graduate students have desks in there. You are filling up notebooks with calculations, talking to future you through your comments in the margins. You are underlining impractically large blocks of text in printouts of peer-reviewed papers and pinning them to a large corkboard. You are grading quizzes and over-preparing for your very short lectures. You're sending lots of emails to the senior scientists who are sitting in their offices one floor below. Sometimes a friend or acquaintance stops by to run a research idea by you. You let them write on your chalkboard, squeezing their equations right by yours.

Most weeks, you attend two seminars and one colloquium. Ironically, you often skip the one that features free coffee and fancy speakers meant to appeal to the whole physics department. It's not that their presentations are uninteresting, but you really want to focus on the obscurities and the nitty-gritty of research that is just like yours, the work of scientists who do science just like you do. When you do get coffee, it is during your self-imposed lunch hour, and you pay for it at the coffee shop tucked into the lobby of the engineering library. As you walk back, you wistfully look over the quad and tell yourself that it's good to be outside. You take pictures of the large, round metal sculpture in the middle of that quad and share them in your Instagram story. It's more palatable than any attempt at communicating your work. And your work is doing science.

All this—the papers, the seminars, the walks—is your world, and you are largely playing by its rules. It's not really a world that was optimized for you, and you are often in rooms where you don't see many people like you. But you chose to be here, to embrace this world as your own—or at least to let it swallow you whole. You rarely feel like a true insider, but after a while the smell of the place sticks to you, gets into your greasily coiffed hair and the scratchy grains of that yellow cardigan. People recognize it at the barbershop and the grocery store: They know you're in town for the university, and that anything else you may like here is incidental. They don't quite squinch their nose, but they sometimes do furrow their brow and tell you that they hated physics in high school. You nervously laugh it all off and retreat to doing the science, which is always waiting to be done.

The number-one rule of your world is that doing science, and doing it well, trumps all. Your older colleagues, mentors, and professors tell you that it doesn't matter who you are if your science is good. If this is really the life for you, you'll never be able to turn your science off. It will always be on your mind, even when you shower, even when you're falling asleep, even when you're on a date.

The phrasing is always clunky. How does one actually "do" science? And what does it mean for science to be theirs? You stumble over it all, many times, but always find a way to pick yourself back up. That's a rule too: you have to have grit, and no setback, sadness, or hurt is ever allowed to defeat your desire to do science. This rule is doubly true if none of your science idols look like you.

You can imagine your future in this world. You know it will take a long time to establish: six or seven years of doctoral work, then three or four of moving around as a short-term postdoctoral researcher, then five years working as a professor who does not yet have tenure but can get there if they just keep grinding. These are also the rules of your world, and you have learned to accept everything that comes with them, such as low wages, and the inevitable transiency of every home you may have in your twenties, and the several years of being separated from your partner, who is trying to follow the same career playbook in another field. But on good days, this future feels so tangible—within reach if you can just keep doing everything that this world expects you to.

You imagine teaching physics classes like those that changed you when you were in college, and approaching that work with kindness. You imagine running your own research group, building it on curiosity and confidence and solidarity. You imagine sitting in a bigger office and reading all those notebooks in which a past version of you asked in the margins about "how to deal with this," then taking a pen to the line next to the comment and dealing with it in a way a mathematically minded person may later call beautiful.

At the end of the day, you're usually working up a sweat, either at the gym or a hot yoga studio. The sense of discipline that you get from doing this even when you are very tired and sad is satisfying. Discipline, you are told, is also something that you need to be a successful scientist. You run laps on the small indoor track on the second floor of the crowded university gym and think about quasicrystals and anyons and topology. You don't really have insights—it's just that your mind stops thinking about your legs and uses that energy to gestate ideas instead. You're not sure if this counts as doing science. Often, you end up getting to your pre-prepped dinner at 10 p.m., eating by the light of a computer screen that glows with the bluish white of the spaces between equations.

Right as you are finishing up your meal, your partner always texts you: "Hey champ, how was your day?" The question is a courtesy because he almost always knows what you've been doing all day. Of course, you have been doing science.

You're not exactly sure what your purpose is here, so you feel like an interloper.

You and your spouse live in the basement, which is divided into a bedroom, a bathroom, a tiny kitchen, and an area where a small couch is trying to manifest a living room. You remember when the bedroom in which you are now sleeping was delineated by three walls and one taut blanket because the room had been victim to a hurricane, and recovery took an awfully long time. You are here because you are, in some sense, also a victim of a natural disaster, and the recovery timeline is currently a complete mystery.

Your bedroom does have four proper walls now, some decorated with vintage cross-stitches, some with photos of someone else's childhood. There are tchotchkes everywhere. You know that many were purchased from thrift stores and antique markets and have no history here, but you still often feel like you have been made to live in an archive of a stranger's memories.

This house is not your house, but it is your family's. And it is your family's house, but not the house of your childhood, because you got folded into this family later in life, through marriage. You remember the first time you visited here and how you could not figure out all the knobs in the shower, so you splashed tap water all over yourself and the person who is now your brother-in-law walked in on the scene. He has since moved out, and you are now an expert in turning on the shower. This time, your and your spouse's visit has no set end date.

Is it really a visit if you feel like you might never leave? You really thought you'd be staying here for just a few weeks, until this virus outbreak subsides, but now months have passed, people are still dying, and this house has become your world.

You still don't fully understand the rules of this world, but you've picked up on some of its hints. One key word is "wife." As a wife, you get to consider yourself an official member of this family. As a wife, you also have responsibilities. You have a hard time with this because you don't always agree on what those responsibilities are. In the past, many of your visits to this house ended in tears of frustration and anger at being asked to put your spouse on a diet, or otherwise whip him into shape, or make him a proper adult man, as if either of you cared about propriety. Now, when you cry during an argument, there is no impending flight or train ride that can put a nonnegotiable end to the argument.

You cook most of your meals most days. You have always loved cooking, but cooking now also serves a different purpose. It is a distraction from the death and fear just outside your door. It is also a way to use up the time that you were forcefully given when all the restaurants and movie theaters closed and your work became remote. And it is a necessity: if you don't cook for yourself, no one else will.

When your mother-in-law cooks, she tells you that she doesn't know how to cook for you, so you will have to join the family dinner table with a dish of your own. Instead of feeling left out, you force yourself to be grateful for the latitude you've been given in the kitchen. This may be a rule of this world, you surmise. Whenever your father-in-law is around for dinner, he tells you how he tells everyone about your cooking. "It's vegan, but good," he says, and you fully believe that he believes he is giving you a compliment. Again, you suspect that the rules say to be nothing but grateful at this moment.

You make tofu marsala, curried lentils of all colors, roasted root vegetables drizzled with rich sauces. You cook sticky stir fries and hearty stews flavored with red wine and plum jam, and you bake myriad, slightly different banana breads. When you're not cooking, the most you can hope for is pasta with tomato sauce or takeout from that one Thai place that manages to make eggplant taste rich despite the economic crisis. A few other restaurants in the neighborhood still offer takeout, but this is the one that you and your family really fixate on. This is the fuel that your world runs on, and you are making peace with that.

Essential to this world is grumbling. Grumbling about the economy, about crime, about immigration, about mayors, about packing the Supreme Court, about women who are too sensitive and people of color who are too angry. Early on in your incursion into this world, you learn to never read the newspapers that your in-laws reserve time and money for. The *New York Post*, it turns out, has a lot to say about people like you.

Before you learn to avoid this paper, it tries to teach you that remote learning is as bad for children as not going to school at all, that teachers who ask about pronouns may as well all be "groomers," and that too many foreigners are allowed to live in this country. In this way you learn a new rule of your world: although you're an immigrant and queer, and although you respect students' pronouns in your classrooms, being a beloved son's wife sets you apart from your counterparts that get reviled in the papers. "Wife" trumps all identity markers that you fear would otherwise make you undesirable or deserving of scorn. When you have the opportunity to co-organize a virtual conference on diversity, inclusion, justice, and advocacy in physics, you make sure to take those calls only in the basement.

Your credentials as a doctor of philosophy, a scientist, and someone who had lived on their own in a foreign country since the age of sixteen don't mean anything under the rules of this world. When you try to speak about imagining a better future, one with more kindness and less competition for artificially scarce resources, you get told that you are naive and don't know anything about how the real world works. *I've been pulling myself by my bootstraps for fourteen years*, you grumble. You are baffled by what this "real world" is or where you may be able to find it.

It's hard for you to imagine any world other than this one at the moment, but you tell yourself that imagination is a muscle that simply has to be exercised. So, you keep trying. In the kitchen, you steam blocks of tempeh so that they lose their bitterness and hope you do not become bitter yourself.

In 1954, Hugh Everett III first considered the idea of there being many worlds while drinking sherry at Princeton.[1] He was a physics graduate student there, out with other physicists, and the story goes that their banter led him to an idea nothing short of heresy. Everett posited that instead of us tripping over oddities of quantum mechanics in our world, it would all make a lot more sense if there were many parallel but independent universes instead. His way of thinking converts an infamously abstract mathematical object—the quantum mechanical wavefunction—into something real. Physicists have so far not found any smoking-gun experimental evidence in favor of his "many-worlds interpretation," but some still ardently support it.

By the time Everett posited his idea, physicists had been working with the wavefunction for several decades, and it was widely accepted as a useful mathematical tool for predicting and explaining most things that happen in the quantum realm. At its simplest, it's a mathematical function that looks like a wave. If you wrote it down on a piece of paper and handed it to a mathematician, they would say, "If this function matches anything physical, then that thing ought to be some sort of a wave." If you turned the wavefunction into a graph and drew it by hand, you would be tracing a shape that rises into crests then dips into valleys. The wavefunction is integral to quantum theory because quantum theory reveals that every object can at times behave like a wave. Accordingly, every object can be matched to a mathematical wavefunction.

Consider, for instance, putting an electron in a very small, inescapable box. If it were a classical object, like a marble, predicting its position at some future point would require you to know things like its mass, energy, current position and speed, the friction at the bottom of the box, and so on. And you'd use equations developed by Isaac Newton in the seventeenth century to make your prediction. But because an electron is a quantum object, not a classical one, what you need to know instead is its wavefunction. The distinction between functions and equations that you use to go from the pages of a lab notebook to the real world may seem like a minute mathematical point, but once you opt for the wavefunction, conventional notions of reality become endangered.

Suppose you have two wavefunctions that represent two different states of this boxed-up electron. One may represent the particle sitting at the center of the box, and another might represent it being closer to the box's edge. If you add the two wavefunctions together, you end up with another wavefunction. This is because waves always combine to make more waves, similar to how small ripples give rise to choppy seas. But what does it mean to add two of the electron's wavefunctions together? Will the electron get choppier?

Wavefunctions formed from the sum of other wavefunctions correspond to "superposition states," and the mathematical structure of quantum theory forces you to interpret them probabilistically. If you know that the electron is described by a wavefunction that is the sum of the "in the center" and "at the edge" wavefunctions, there are no formulas that can help you determine where exactly the particle is. What you can calculate from this wavefunction is the set of probabilities of finding the electron at one of the two positions if you open the box.

One interpretation of this is that the electron is in both places at once. Though this interpretation is common in popular science books, and some physicists use it as a shorthand way to explain the phenomenon, it is not the most accurate—it invites unphysical mental images like the particle breaking in two or somehow doubling itself. On the basis of decades of experiments, researchers agree that neither of those two things happens. A more accurate way to imagine the electron in the superposition is to picture it as a fuzzy, extended cloud of possibility. In the previous example, the cloud would envelop both the center and the edge of the box. The way to end this fuzziness and make the electron definitively exist in one place only is by opening the box and measuring its position directly.

Once you do that and measure the electron's location—for instance, at the center of the box—all the other probabilities and options somehow go away. Werner Heisenberg, one of the founders of quantum theory, wrote, "The transition from the 'possible' to the 'actual' takes place during the act of observation."[2] After observing the electron, you know exactly where it is, and you know this with certainty. The only equation you need from now on is the "in the center" wavefunction—you can completely forget about the other one. The act of observation seems to save the electron from being a hard-to-imagine chunk of probabilistic fuzz.

Heisenberg theorized this "wavefunction collapse" in 1927.[3] In the intervening years, the wavefunction has become less esoteric and abstract—physicists have found ways to identify its more physical manifestations. For instance, in 2021, researchers effectively measured the wavefunctions of quantum particles of light in the lab.[4] This breakthrough transformed the wavefunction from being a formula that lives on the page to an experimentally accessible quantity haunting the displays of measuring devices. Wavefunction collapse, however, remains mysterious.

Physicists of all stripes acknowledge that something like wavefunction collapse must happen in their theories and experiments, but they have not yet agreed on a detailed or intuitive picture of how. If the electron in the unopened box has, for instance, a 50 percent chance of being in the center and a 50 percent chance of being at the edge, what exactly is it that pushes it into one position or another when the box is opened? In 2024, several Nobel laureates released a compendium of the most important questions facing contemporary physics. They explicitly list this question: "How does a quantum measurement decide which outcome is observed?" The compendium places it at the same level of importance as the question of whether our inventory of every particle in the universe is complete.[5]

A philosopher told me in 2024, "People should be thinking about this if they're interested in the nature of reality."[6]

Some physicists argue that this phenomenon simply illustrates that an observer, or experimenter, can never be fully removed from the system on which they are experimenting, so they must also always change it. Others will tell you that there simply are fundamental limits to how much we can know about physical systems. Heisenberg suggests that, in this view, the collapse of the wavefunction teaches us not just about nature, but our methods of studying it. He writes, "We have to remember that what we observe is not nature in itself but nature exposed to our method of questioning."[7]

Everett's theory puts an end to all such philosophizing because it fully eliminates wavefunction collapse from quantum mechanics. In his many-worlds interpretation, everything about the wavefunction is real, and all the probabilities that come attached to a superposition state also refer to something real. In Everett's view, if there is even a 0.000000001 percent (or arbitrarily smaller) probability of finding the electron at the left edge of the box, there is some reality in which the electron is indeed at the left edge of the box. This reality is one of the many worlds.

When you perform a measurement, the outcome you find is the outcome that matches the world you live in. Simultaneously, copies of you exist in myriad other worlds where other outcomes are true, and your copies experience them. Every interaction between a classical observer, like you, and a quantum object, like an electron, creates new branches of the multiverse—new parallel worlds.

Though this interpretation of quantum theory renders moot some of its more philosophically thorny features, it is not without its own problems. Primarily, it complicates the idea of being a person. Everett himself seems to have understood that there was something about his theory that was troublesome for the notion of selfhood. His doctoral thesis—the first draft of the many-worlds interpretation—was published in an abridged version in a journal, but in an earlier draft he illustrated the way that the wavefunction of everything branches by invoking an ever-dividing amoeba: "One can imagine an intelligent amoeba with a good memory. As time progresses, the amoeba is constantly splitting, each time the resulting amoebas having the same memories as the parent. Our amoeba hence does not have a life line, but a life tree."[8] Everett's PhD adviser, John Archibald Wheeler, prompted him to omit the language of splitting—there was no way to visualize what would happen to a splitting person that was not off-putting.

In the journal article, which was Everett's only academic publication, he shows only slightly more warmth when discussing the role of observers, writing that "as models for observers we can, if we wish, consider automatically functioning machines, possessing sensory apparatus and coupled to recording devices capable of registering past sensory data and machine configurations."[9] Here, the necessary conditions for being an observer are simple enough to be satisfied by a machine. This, however, leaves room for many questions about what a conscious person, and not just an automaton, could make of their multiversal existence. As philosopher Emily Qureshi-Hurst writes, some six decades later, "What are we to make of the idea that the human person ought to be thought of like a phylogenetic tree, a set of descendants branching off from each other forever but sharing a common ancestor? The philosophical implications for our understanding of personal identity are mind-boggling."[10]

There seems to be a fundamental contradiction. On the one hand, each copy of you in each of the many worlds is bound to believe that they are the real you. However, the real you should be described by the universal wavefunction that adds all of them together, implying that each parallel you is only one part of some larger, universal you.[11] The problem only gets worse when you turn to the question of consciousness. When a person shifts into a new branch of their multiversal self, does the branching happen slowly enough for their neurons to fire and spark consciousness? Scientists continue to argue about this.

In 1959, Everett met with Niels Bohr, who takes credit for the most conventional interpretation of quantum mechanics but was dismissive of Everett's work, which the younger man found discouraging, describing the meeting as "doomed." Everett never published any more work on quantum theory. He found employment at the Pentagon, where he mostly developed mathematical models for the Weapons Systems Evaluation Group, much of which remains classified.[12] His models helped establish the concept of "mutually assured destruction" in a potential nuclear war. Later, he worked on similar defense research as a contractor. He unexpectedly died of a heart attack in 1982, when he was fifty-one and rather disconnected from his old life as a physicist.[13] Everett was not the only scientist to think of the multiverse—Leon Cooper, Deborah Van Vechten, and H. Dieter Zeh also developed similar interpretations of quantum theory—but in 1970, an article in *Physics Today* by physicist Bryce DeWitt heavily focused on Everett's paper and his role in developing the concept. In the article, DeWitt coined the "many worlds" moniker, propelling the theory to a level of fame that continues to echo in both science and science fiction today.[14]

Current versions of the theory provide more detailed explanations for why researchers may think they're seeing wavefunction collapse, but none have yet offered a surefire way to observe the splitting of worlds. Today, wavefunction collapse is often explained by the notion of "quantum decoherence," a sort of degradation of quantumness that follows from objects connecting with their environment, which is an idea complex enough that it became its own area of study.[15] But interpretations of quantum physics built on the notion of decoherence are still philosophically complex—and can still be hostile to the idea that a human observer is at all necessary. In 2019, physicist Sean Carroll wrote, "Everett was right; once the whole Universe is your subject of study, it doesn't make much sense to carve out a special place for a classical observer."[16]

In 2024, I interviewed a group of physicists who had developed a theory that contained even more worlds than Everett's—whole realms of new worlds, in fact. They argue that our world belongs to the realm of worlds in which there is an intuitive split between us as observers and the quantum objects that we observe. But many more realms of worlds exist in which we are not important for that split at all and it is interactions that do not have to respect the idea of a whole, coherent person that give rise to new multiverse branches. For instance, it could be the interaction between your left elbow and some distant part of the Andromeda galaxy that gives rise to a new branch while completely ignoring the fact that the rest of you is attached to your elbow too. In these parts of the multiverse, the distinction between the observer and the observed absolutely does not care about our consciousness or selfhood.[17] The researchers told me as much when I interviewed them: part of their motivation, they said, was to "get rid of the human perspective."[18]

Everett would probably disagree. He believed that the existence of many worlds would grant him immortality. Consider, again, the electron in the box that is in a superposition state of being in the center and at the edge. Suppose that every time you open the box and find it in the center, you immediately smash the electron into pieces. Following Everett's line of argument would be analogous to saying that that's not a problem, because the multiverse guarantees that there is always a parallel universe in which the electron lives on at the edge of the box. He believed that his selfhood was a giant superposition of multiversal selves, and that in the instance of any adverse effect—the equivalent of being smashed—his consciousness would find a way to shift into a branch of the multiverse where he continues to live. According to most accounts, in the branch that he shared with his biographers and family members, he was a bitter man who drank often and failed to build relationships.[19] It is hard to not wonder whether he thought some of his multiversal selves had done better.

The many-worlds interpretation invites this sort of drama—it seems to leave you with a choice of immortality or nothingness. In this way, it reflects some of the anxieties of living in our world, where a person can briefly feel extremely important, then walk into a different room and feel incredibly small. Or at least I can.

In 2014, science writer Rowan Hooper discussed the notion of morality in Everett's multiverse with several experts in the pages of *New Scientist*.[20] Ignoring all the questions of what being a self actually means within the many-worlds scenario, the piece warns the reader: "Every decision you make may spawn parallel universes where people are suffering because of your choice." Over the course of several thousand words, Hooper's worry for the selves he may never be able to meet feels rather real, and prominent Everettians earnestly engage him on it.

Physicist David Deutsch tells him, "If there is a small chance of an adverse consequence, say someone being killed, it seems on the face of it that we have to take into account the fact that in reality someone will be killed, if only in another universe."[21] Writing a few years later, Qureshi-Hurst is more direct about the possibly troubling fates of the people that she will branch into. "Out of all these people, there must be at least one who is living the worst possible version of my life, and there will be countless more living very bad lives. In other words, there is much more suffering in the Everettian multiverse than there is in a singular universe," she writes.[22] Qureshi-Hurst does acknowledge that many versions of her would also be living very good lives but, in her view, suffering and joy cannot be put on the same footing. The abundance of suffering in the multiverse cannot be simply canceled out in this way because, she argues, the pain of her equivalents in unhappy branches is simply not worth it.

Hooper, for his part, ends his piece by siding with the physicist Max Tegmark, whose perspective has a somewhat more positive spin. Tegmark shares that if he finds himself in a dangerous situation, like reckless driving, that could cause the death of one of his parallel-universe replicas, then "the minimum tribute I can pay to that dead Max is to really think through what happened and learn some lessons."[23] Hooper admits to the reader that there may not be one definitive answer regarding how to live morally when there are many worlds, but he asserts that their possible existence can, and maybe should, "help us think harder about our choices and actions."[24]

I huffed and puffed at this at my own desk in the *New Scientist* newsroom when I first read it. I had come across it while writing about the theory of the many more realms of worlds, which felt rather attractive in its discounting of the observer. I wanted to side with those researchers and take the human observer out of the picture. I wanted the multiverse to tell me that I don't matter all that much. This would have offered me some relief. Even if the different worlds that you move through are more metaphorical than Everett's literal multiverse, it is very tiring to have to keep trying to make sense of them, and what you're supposed to do within them.

After spending years in the world of physics graduate study, and after that seemingly endless year in which my world was reduced to the basement of my in-laws' home, that sense of tiredness settled into my body with a stunning heaviness. In the intervening years I entered more worlds, simultaneously thinking that I was blissfully unimportant to the multiverse and also dreaming of a decision that could land me in one of its happiest branches. The two tugged at me in opposite directions, neither of them pleasant. Thinking that I was unimportant, barely a coherent entity, gave me permission to wallow and reject connections with people who love me. Thinking that I was one decision away from the multiverse's happiest branch made me stressed and anxious. I continued to be tired. I still might be.

As a physicist, what should matter to me most about the many-worlds interpretation of quantum mechanics is that it is incredibly difficult to test experimentally. In fact, it is quite possible that there will never be an experiment that will serve as an unambiguous test for the existence of the multiverse. A researcher once told me that the many-worlds interpretation could be correct, but that this may also not matter. Perhaps we have all been living in the multiverse this whole time, without ever noticing that we are always becoming myriads of people. Even through the fog of personhood fatigue, however, it is clear to me that there is something worth holding onto in the idea that quantum theory, our most rigorously correct theory of the world, allows you to think that every single one of your options for being is perfectly real.

You are here to bake cakes, to wash grains of rice in your sink until water runs clear every time, to cut soft tofu into jiggly cubes begging for spice, to slice apples and dip them into peanut butter, to boil pasta perfectly al dente every time.

No.

You are here to write stories, to talk to many people about many things and let their words filter through you and onto the page, to find just the right combination of words that can make someone go "wow," that can make someone look at the world with just a little more awe.

No.

You are here to put miles on your favorite pair of running shoes, to watch your chest glimmer with sweat in the dirty gym mirror, to lift something heavy, to feel the tension in your shoulders and know it will soon become strength, to trust your thighs and your arms and the breath that keeps you together as you move from one precarious posture to another.

No.

You are here to say "I love you" to a lot of people and mean it every time, to text countless good mornings and good nights, to share meals that cost anywhere between $5 and $500, to play that one board game with the birds even though you still don't really know the rules, to go dancing even though your feet only ever move awkwardly, to offer a partner thick slices of homemade bread slathered in a relish that tastes like your childhood on a late sunny morning, to ask for small kisses, and maybe big ones, without shame.

No.

You are here to do all of it. You are here to do even more. You are here to be even more, to be every you possible. There are many worlds, and they are all yours.

CHAPTER 5

SCHRÖDINGER'S CAT

I'll paint the scene for you as vividly as I can: a middle-aged woman in a deerstalker hat is holding a magnifying glass and loudly speaking German. Behind her, on the rounded wall of the hotel ballroom, is a fuzzily projected image of a man standing by a lake. He is the German physicist Erwin Schrödinger, the woman is reciting from his writings, and this is a play about the intricacies of quantum physics.

I, the production manager, am nervously standing in the shadows, waiting for the woman to hit my favorite phrase in her monologue: "*gemischt oder verschmiert sind*."[1] She gestures with her hands, miming something like spreading butter on bread. The hard *sh*, *g*, and *t* sounds roll over her lips with vigor and conviction. There are no subtitles provided, but the audience—most of which, like me, does not speak German—is rapt.

The monologue is an excerpt from Schrödinger's 1935 paper, "The Present Situation in Quantum Mechanics," in which he introduces the thought experiment that is now known as "Schrödinger's cat." He describes the scenario as a "quite ridiculous case"—even "burlesque" in some translations—but presents it anyway.[2]

A cat is confined in a closed metal chamber that also contains a "diabolical device" that the cat cannot interfere with.[3] The device is diabolical because it's filled with a tiny bit of a radioactive substance that has a 50 percent chance of decaying within an hour, and if it does, the decay will trigger the release of a poison and kill the cat. After an hour in the chamber, the cat has a 50 percent chance of being dead and an equal chance of being alive.

Schrödinger applies the rules of quantum mechanics to this cat, which leaves him with a conclusion that he deems so ridiculous and unsavory that he apologizes to the reader. "Pardon the expression," he writes, while describing how quantum mechanics would render the living cat and the dead cat "mixed or smeared out in equal parts," or, in the original German, *gemischt oder verschmiert*—my favorite line in that play.[4]

How do you make sense of a cat that is dead and alive in equal parts? You open the box and take a look. Once you do that, nothing is "smeared out" anymore; the cat collapses into just one state, either unambiguously alive or unambiguously not.

I moved to the United States in August of 2008, just as my junior year of high school was about to start. Coming from Croatia's biggest island—my ideas about American culture and politics almost fully based on reruns of *The Fresh Prince of Bel Air*, various *CSI: Crime Scene Investigation* shows, and one Rage Against the Machine record—I arrived in upstate New York, where I would attend an expensive and isolated private boarding school for girls. My classmates were wearing Tory Burch flats, carrying Vera Bradley bags, watching *Gossip Girl*, and humming along to early Taylor Swift. I was wearing oversized Iron Maiden shirts and Converse sneakers, rambling on about string theory. I was visibly foreign, less wealthy than most, and way more clueless. No one really knew where Croatia was, and my Slavic accent, island-dwelling tan, and Italian-sounding first name didn't make me any easier to classify. I confused people simply by being there. And I bored them by always talking about "back home."

Even in my Croatian high school, I had been an odd, depressed kid with niche interests. Now I was lonely in a way that could not be remedied by finding another student who had the same favorite book. And I was gaining so much knowledge about what existed beyond my hometown's worldview that it felt like my head was perpetually spinning.

Having always been a nerd and a try-hard, I did well in my classes. Eventually I also managed to form fragile friendships with some of the other recent transfer students. Nonetheless, it all continued to feel surreal.

It was what William Gibson calls the "mirror world," where "things are just like home—but not quite."[5] And I could not wait to go home.

I finally landed back in Croatia during that first winter break. Even though I only had three months of American life under my belt, my family could tell, they said, that my time abroad had changed me. They declared me virtually American. I have felt every possible feeling about this, from sadness to rage. But my relatives were not incorrect: in December of 2008 I had already become split between two places, my personhood smeared across four thousand miles. Like Schrödinger and his cat, I was unsettled by this state of being—and would continue to be for years.

My mom's aunt says I roll my *r*'s softly now, unlike a true Croat. "You speak so quietly now," my grandfather says. "That's not what we're like." When faced with this recognition that my speech has changed, my mixed state, American and not-American at once, cannot hold. I collapse into American-ness, my tone, intonation, and volume all wrong.

At boarding school, my class reads *The Great Gatsby* and *Death of a Salesman*, and I have no idea what the American Dream is even though I am supposed to be living it. I get a B in English, a rarity in my lifetime of anxious overachieving. My mom is baffled. "Your English has always been so good," she says. I collapse, fluently.

Schrödinger's "cat paper" became the most famous of his career, rivaled only by the earlier one in which he introduced what is now known as the Schrödinger equation. This equation is an indispensable mathematical tool for all of quantum physics. The function that solves it sparked countless arguments among the great physicists of the time—some of which, in turn, inspired Schrödinger's grotesque thought experiment.

In *The Age of Entanglement: When Quantum Physics Was Reborn*, Louisa Gilder gives a fanciful account of how Schrödinger arrived at his famous equation.[6] She also sanitizes him, at best glossing over his habit of having affairs with underage students and at worst romanticizing one of those instances. When I read her book at age sixteen, the disturbing nature of those affairs did not make an impact on me, and I would only realize how much they darken Schrödinger's legacy more than a decade later, when I considered recommending *The Age of Entanglement* to one of my students. I think I had ignored this grim personal detail because I'd been enthralled with its physics. Gilder's book, which reconstructs and reimagines many conversations between storied physicists, successfully convinced me that there is nothing more romantic in this world than being a theoretical physicist. My romance with the profession would be tested many times later, but the tale of Schrödinger's equation maintained its place in my memory.

It starts in 1922, when he contracted tuberculosis and had to retreat to a health resort in the Alps. While he was convalescing, the work of Hermann Weyl, whose mathematical insights undergird much of modern physics, inspired him to think of the electrons inside of atoms as tiny standing waves, like those sustained on a plucked guitar string but smaller and more circular.

What happened next is often chalked up to a 1925 comment presumably made by Peter Debye, who was at the time a decade away from winning his Nobel Prize in Chemistry. Though Debye later denied it, the story goes that after seeing Schrödinger speak at a colloquium, Debye told him that a serious physicist would not be thinking about the electrons' wave character without first deriving a formula called the "wave equation."[7] Wave equations already existed for all other types of waves in physics, ranging from water waves to electromagnetic waves. In fact, to call something a wave, physicists often required that it make sense after being plugged into a wave equation. Regardless of whether it was Debye that spurred Schrödinger to think about this missing piece of his physics or not, Schrödinger and a lover returned to the Alps that winter, where, in Gilder's telling, he worked through Christmas "with pearls in his ears to dampen any distracting outside noise." Eventually, he descended from the snowy slopes with a wave equation in hand.[8]

The world of quantum physics was small at the time, so it was Weyl who then helped Schrödinger solve that equation. Keeping the teenage me at the edge of their seat, Gilder limits her explanation of the pair's calculations to a brief parenthetical: "A solution to the Schrödinger equation, describing the state or condition of a given quantum entity, is known as a wavefunction."[9] She finds more space for the lyrical assertion that "Schrödinger was beginning to believe, looking down on the undulating whiteness all around, that the entire world was made of waves."[10]

Teenage me had no idea what a wavefunction was, but I could understand the boldness of this statement because nothing that surrounded me—the desk I was sitting at, the book I was holding—felt remotely wavy. Yet, this is what the quantum character of the world boils down to: any quantum object is actually a wave, and it is described mathematically by the wavefunction. You know something is a meaningful wavefunction—that it can teach you something about the real world—because it solves Schrödinger's equation.

When you first start learning quantum mechanics, your days become consumed by solving different versions of Schrödinger's equation and finding corresponding wavefunctions—like for electrons inside of atoms, or for electrons stuck in narrow, infinitely deep holes. Once you have a wavefunction, you can subject it to different mathematical operations that then produce measurable quantities about the quantum object to which the wavefunction corresponds. Do you want to know what the object's energy is? Or where in space it might be? Or where it will be in six hours? You can try to use mathematics to coax this information from the wavefunction.

This utility of the wavefunction has been confirmed again and again, from the earliest experimental tests to its applications in contemporary electronics. But the question of its interpretation—what exactly the wavefunction means before you've done anything to the object that it represents, mathematically or in the lab—became a contentious topic among Schrödinger and his contemporaries, including Albert Einstein, Niels Bohr, and Werner Heisenberg. It is still mysterious today.

The key question is illustrated by the thought experiment with the cat. Schrödinger's equation tells you that the "dead cat" wavefunction can be added to the "living cat" wavefunction to create a perfectly valid state of the cat. That is, if the living cat is one solution of Schrödinger's equation and the dead cat is another, a mix of the dead and living states must also be a valid answer. Therefore, you must take this mixed state seriously and think of it as physically real. In the 1950s, Hugh Everett III used this as a starting point for asserting the existence of the multiverse, but the more orthodox interpretation was formulated twenty years earlier, through the work of the physicist Max Born.[11]

Born was studying how quantum particles collide when he first thought of interpreting mixed quantum states probabilistically. This both enraged physicists of his time and also really stuck with them. Within this framework, now known as the Copenhagen interpretation, a wavefunction that is a superposition of two others means that the corresponding quantum object is potentially in one of those two states—but without interacting with it, without forcing its wavefunction to collapse, the best you can do is determine the probability of one of those states being true. In this view, Schrödinger's cat has a 50 percent chance of being dead and a 50 percent chance of being alive—that's all you can say without opening its chamber. You may imagine it as "smeared out" within the box, but you really cannot know anything other than those two probabilities. They, in turn, reflect the details of the setup – you could change these probabilities, for instance making the cat's probability of appearing alive only 30 percent of the time, by building the diabolical device that traps it differently.

Schrödinger always hated the fact that quantum theory could not commit to a more definitive account of the cat's reality. Why did you have to open the chamber to know what's up with the cat? Did objects as mundane as cats not exist in an un-mixed, un-smeared state unless someone was looking at them? Schrödinger and Einstein wrote to each other about this repeatedly, often expressing frustration with other physicists' accepting probability as a key feature of quantum theory. In 1960, Schrödinger wrote a stern letter to Born, berating him for planting the first seeds of the probabilistic interpretation. "Have you no anxiety about the verdict of history?" he asked the younger man.[12] Schrödinger died a year later.

By the time my path intersected with quantum physics, history had seemingly made its verdict. In the 2010s, I learned a version of quantum mechanics that relied on the Copenhagen interpretation, and I memorized mathematical rules named after Born. The first proper quantum mechanics textbook that I read—the mathematically challenging tome by Ramamurti Shankar with the red cover—contained a numbered list of foundational postulates of quantum mechanics.[13] The third entry on this list was a statement that all you can know about a quantum object before you collapse its wavefunction is the set of probabilities governing the object's properties, such as its location. Its fourth entry? A strict requirement that you get that wavefunction by solving Schrödinger's wave equation.[14] I don't think he would have approved of his work being presented right alongside the "reigning doctrine" that he found so preposterous.[15]

When college friends take me out to dinner, I have to learn how to eat noodles with chopsticks, how to fold a slice of pizza, and that you eat Ethiopian food with your hands. I collapse, with sticky, clunky fingers.

On a video call, I can't remember the Croatian word for "condition," so I just say it in a Croatian accent. Later, it's "transportation" that gets me. On the other side of the screen, my father is furrowing his forehead. I collapse, tongue-twistedly.

As a college student at the University of Chicago, I don't understand why you have to order fried chicken through bulletproof glass at an iconic hole-in-the-wall joint. I don't know what "deli" and "bodega" mean when I visit New York. I collapse, hungrily.

By the time I was reading Shankar, I had been splitting my time between the United States and Croatia for almost five years. My trips home were costly and emotionally heavy, but at least I rarely got into legal trouble with immigration. In 2010, when I was preparing to go to college, by a stroke of luck I was granted a student visa that would remain valid for four years, or just enough time to complete my bachelor's degree. Going to college in Chicago felt so freeing after the strict, confining rules of my boarding school, but there were always strings attached to my immigration status. On forms that required me to put myself in one of three categories—"citizen," "permanent resident," or "other"—I checked the third, and I tried my best to keep track of all the paperwork.

When I went "back home," I had to bring extra papers to the airport, and some had to be freshly signed by an administrator at my school. Once, when I was returning to college from Croatia, this signature was several days past its expiration date and I ended up in a back room of Chicago O'Hare Airport, stuck in a limbo I hadn't even known existed. Immigration officers kept trying to call my university to confirm who I was, but it was Sunday, so no one was there to answer. When I finally made it back to campus, the American boyfriend with whom I was living was bewildered first by my lateness, then the notion that an immigration issue had delayed me. He had assumed that the system worked better—or that it at least worked better for educated white Europeans like me.

Throughout my college years, whenever similar issues cropped up, Americans around me expressed a kind of benevolent outrage and confusion that was also quite inert. I was immensely lucky to have received oodles of financial aid from a private family foundation, something that tends to be nearly impossible for international students, but money problems still always loomed. When I explained to my research mentors that I was not eligible for any of the commonly recommended fellowships because I was a foreign national, they'd briefly conjure up a somewhat empathetic emotional response, then suggest that I just work for them for free. I coauthored several papers in college, all based on work I was never paid for, and all while my taxes contributed to some of the fellowships that helped my American peers get paid for similar work or spend summers learning new skills and racking up more varied research experience.

When I started working as a teaching assistant in my second year of college, my job orientation was mostly just this: don't sleep with a student, don't sleep with the professor, and know that you will not get paid for a minute longer than the puny number of work hours that your visa allows. When recruiters for industry jobs and teaching nonprofits came to campus, they'd sidestep questions about international applicants by obliquely referring to a labyrinthine system of extending one's legal stay in the United States through training programs. Later, I would run into older international students from my college who had at some point gotten into legal trouble for inadvertently breaking the rules of such programs. Once, a Teach for America recruiter put me through an hour-long interview before realizing that I was not actually eligible for the position she'd been buttering me up for. Her eyes glazed over with frustrated exasperation when that realization finally hit.

Being an immigrant always made me tired because a strong sense of temporariness kept looming over me, but I was lucky that this was the extent of my troubles. My financial situation was never comfortable, but I could afford off-campus housing, and when I had to fly home to renew a visa, I could distribute the associated costs across credit cards and savings in a way that avoided long-term devastation. My peers from the Middle East and Asia had visa stories that were far more harrowing: repeatedly rejected applications, or application approvals that permitted only one or two years of education in the US at a time. Even professors who were not American citizens encountered problems when trying to travel to conferences in American cities if their research program featured terms like "atomic" or "nuclear," both of which are ubiquitous in contemporary physics. And the gossip about international students from India, China, or South Korea—all significant populations in any physics department—was always tinged with a kind of racism and colorism that I know I will never experience. It speaks to the remarkable resilience and drive of international students that, most years, American universities award them almost as many PhDs in disciplines like physics as they do to those that hold an American passport.[16]

By the time I was taking my first quantum mechanics class, I felt that I had a clear binary choice: to either find a way to become an American citizen, or to return to Croatia once I'd run out of degrees I could pursue. With some shame, but almost no doubt at all, I chose the former.

A boyfriend takes me on a date to the Cheesecake Factory and, years later, someone tells me that this is not a romantic venue. It's sort of grotesque and embarrassing, I learn. I collapse, heartbrokenly.

I joke about my Southern college boyfriend being "white trash" because I don't understand the severity of the term. He refuses to speak to me for a whole day. I collapse, offensively.

My father's cousin asks whether the crisp Croatian air hurts my New York City lungs. I collapse, breathing heavily with grief.

Though I have spent years working toward becoming American, something about it has always felt unsavory. I have never been ready to give up the life I've built for myself here, but I worried—and still worry—about implicitly endorsing American culture as somehow superior, or accidentally falling into its most consumerist, capitalist habits. Studying the history and politics of the United States has taught me that pursuing an American identity means throwing in my lot with institutions that often have racism at their foundations, and propping up a generations-long project of nationalism and colonialism. The green card that I now carry in my wallet often feels like a badge of complicity. It reminds me that I saw the flaws of the United States, then nonetheless begged it to fold me into its melting pot.

But the prospect of returning to Croatia, and then conforming to the standards expected of a Croatian adult, has long felt worse than shameful or undignified—it has felt virtually impossible. I get a glimpse of it when I visit my family: how unusual it is for someone my age to not have a child and own property, how homophobic jokes still elicit laughter, and the way going to church on Sunday mornings is a given. About 80 percent of Croats are Catholic, and it is not unusual to see the crucifix in public spaces, like the offices of a city's government. Many schools all but mandate that students attend Catholic religion classes that teach the kind of conservative family values that today are peddled by the American right. When I ask my mom, a wildly progressive person by Croatian standards, whether there are queer celebrities in Croatia, she can't name a single one. At sixty, she is still arguing with my grandparents about whether it is embarrassing for her to rent an apartment, be single, and wear summer dresses to work. Her one queer friend tells me they wish they had been born straight—and that the culture of the place where they live, a city with a tradition of punk and political rebellion, still does not feel conducive to them ever coming out. It's not that I have never heard similar stories in America, but the total population of Croatia is about 1 percent that of the United States, so there are simply fewer places to carve out a refuge. The kind of conservatism that may be associated with just a swath of US states is, for the most part, the dominant ideology in Croatia.

I am certain that there are pockets of Croatian society where queerness flourishes anyway, but the prospect of looking for them, likely against my family's expectations, feels heavy and intimidating. In the US, because of my privilege as a white person with an advanced degree and the ability to live in New York City, I have been able to carve out a bit of space for myself, affording me the kind of community and freedom that I cannot even imagine existing "back home." I have never been a legal "other" in Croatia, but an emotional and cultural otherness exudes from me whenever I return there. And the emotional connection between the sixteen-year-old me of my family's memories and the person I am now is thinner and flimsier than gossamer.

That sixteen-year-old was a nerdy girl who liked heavy metal, unquestioningly adored her dad, kept her curly hair long and big, and ate sheep cheese and her grandfather's sausages with gusto. She gave way to someone who looked like a frumpy American tourist, cycled through every stage of feminist theory in rapid succession, refused to eat anything that may have once caused an animal harm, cut their hair, and spoke of gender in a way that the Croatian language doesn't even have words for. Most of the people who knew me when I was sixteen would not be able to explain what I went to school for or why it took so long, even if they've kept track of me, even if we're related. And they have long stopped being able to make sense of me, even when I've hidden the parts of me that seemed most different and foreign. After years of plotting and rehearsing in my mind, I gave up on the idea of ever coming out to my father as anything (bisexual) or anyone (nonbinary). Even with my straight, cisgendered, male American husband by my side, I was already too much.

Ironically, many of the identities and political views I hold are not fully mainstream in the United States, either. Plenty of Americans are offended by veganism (as my father-in-law once joked, most American men don't greet their spouse with "Wife, make me a vegan dinner"), leftist politics, and queerness. In fact, recent years have seen a record-breaking number of legislative efforts to, if not eradicate queer people like me, at least push us into real material hardship by making gender-affirming care less accessible, loosening protections against discrimination and—as started happening in 2025—making it difficult to obtain essential documents such as passports. In 2024, more than five hundred bills targeting the rights of trans people were introduced across the United States.[17] Even as an adult living in a liberal bubble, my life in the US would have probably been easier had I pushed down some of the more non-normative aspects of who I am. Unfortunately, it was the promise of freedom, and the sporadic small tastes of it, that stuck with me instead.

My parents have always considered themselves open-minded and willing to go against the grain of mainstream culture, but that is only true compared with Croatian conservatism. For them, America was distant enough and debauched enough to catalyze my departure from our traditional cultural values, which continue to be heavily shaped by Catholicism and nationalism. While the land of the free was placing constraints on how I could materially improve my life, the view back home was that too much American freedom had turned me into a stranger. When I visited for the holidays, the specter of America was always with me at the family dinner table. My uncles, cousins, and grandparents found it incredibly useful to allude to it when they disagreed with me or wanted to invalidate my beliefs and feelings.

In the nineteenth and twentieth centuries, many Croatians from the island of Krk, where I grew up, moved to the United States in big migration waves. Many settled in New York, and about twenty-four thousand Croats still reside there.[18] My grandfather remembers relatives who made that journey when he was young. They sent him green dollars and orange cheese and became characters in our family folklore, even though we lost touch with all of them and details of their stories have mostly been forgotten. The problem with me is that the details of my story are here for all to see. And they are all mixed up and confused, in dire need of being collapsed into something more sensibly simple.

Unlike Schrödinger's cat, which stays in a single state once you have collapsed its wavefunction, I could never stay collapsed for very long. I would always return as a different kind of mix smeared way outside the edges that had been drawn for me, anxiously fearing that I'll be pushed to collapse again, and again, and again.

My Croatian accent never goes unnoticed at dinner parties, and people always ask where it's from—either shyly, like they're whispering something dirty, or brashly, like I owe them a TED Talk. I collapse, feigning calm.

An American man asks me to speak to him in Croatian during sex. I collapse, my body screaming for help.

When my father remarries, my whole family chips in for a honeymoon, but no one tells me this or reaches out to me for a contribution. I only learn about it way after the fact. Doing the math in my head quickly reveals that a modest fraction of my American paycheck could have paid for all of it. On my next visit home, my father goes on about how great the honeymoon bed-and-breakfast was, what a luxury. I collapse, lonesomely.

A graduate school professor asks me, on the first day of class, in front of everyone, whether there are still snipers on the streets of Croatia, twenty years after the end of our Independence War. I collapse, nonviolently.

Almost a hundred years after its conception, Schrödinger's cat sits in a somewhat paradoxical place. On the one hand, in laboratories around the world, many quantum objects have been successfully prepared in superposition states and observed to "collapse" into a single state, which has scrubbed quantum theory of some of that early weirdness and grotesquery. While the mechanism of collapse itself is still considered to be a mystery, it is much rarer to encounter phrasing like "smeared out" in academic publications. Physicist and YouTuber Sabine Hossenfelder writes that "a superposition is just a sum of wavefunctions for particles that are in two definite states. Yes, it's just a sum." There is nothing weird about this, she argues in an accompanying video.[19]

Physics professors today often explain that Schrödinger's cat is a teachable thought experiment but not an accurate scenario, because we now know that an object the size of a cat can never be truly quantum. Quantum physics works exceptionally well for predicting the behavior of objects that are very small, but as they get larger and larger, classical physics takes over. If you try to calculate the magnitude of any effect that would reveal the quantum nature of a cat-sized object, the numbers will always be so small that they are effectively impossible to measure. In fact, the record for the largest object put into a quantum superposition state was broken in 2023 by a sapphire crystal that weighed about a microgram.[20]

So, there is something like resolution: we know small objects can exist in superposition states because we have seen this in experiments, and for large objects, the effect may never be large enough to make a difference. This unfussy line of thinking, along with changes in the models for funding and professionalizing physics, has reoriented physicists toward using superposition states instead of pondering their meaning. For instance, something called the "cat state" is now a standard ingredient for building and testing quantum computers.[21] Also known as the Greenberger–Horne–Zeilinger (GHZ) state, it involves a superposition of two possible behaviors of a quantum bit—the quantum version of a conventional computer bit—that seem like they should be as mutually exclusive as life and death.[22]

In 2024, I reported a story about what was then the largest GHZ state ever created. It had been done inside a quantum computer.[23] The details of the experiments were somewhat overwhelming, but as I waded through them, I was struck by the language the researchers were using in our correspondence, casually referring to their accomplishment as the "fattest Schrödinger's cat."[24] Another researcher showed me his list of all the notable cat states that have been made in recent years. He had been keeping it simply out of curiosity, which underscored for me just how run-of-the-mill this quantum phenomenon can be to a physicist.[25] By April of 2025, the record for the largest Schrödinger's cat of this sort had been broken two more times. And then there's the French quantum computing start-up Alice & Bob, which is hoping to beat out computing giants like Microsoft and IBM by going all in on machines that contain a microscopic version of Schrödinger's cat, made from light stuck inside special chips.[26] There is nothing burlesque about these "cats," and hundreds of millions of dollars have been invested in them.[27]

Physicists have also learned how to obtain at least some information about quantum objects without dramatically affecting their states. Capturing everything you may want to know about an electron or a photon without collapsing its wavefunction is still considered impossible, but in several experiments, what are known as "weak values"—which change the wavefunction minimally in exchange for some tiny amount of objectively real information about the particle—have been implemented. Often, these are very intricate procedures involving quantum light, a photon playing the role of the cat as it travels through lenses, mirrors, and crystals. Weak values were first theorized in the late 1980s, but it took almost twenty-five years for physicists to figure out how to take these quickest, dimmest, and blurriest of peeks at the catlike objects without the poor smeared-out things realizing they had to collapse.[28] Some of these experiments have been received with skepticism, but they continue to be an active area of research—and are possibly some of our best tools to get as close to clarity as the quantum world will allow.[29]

The longer I spent living abroad, caught in a state of being both American and Croatian but also neither, the more strategies I developed to minimize the effect of someone seeing me as being only one of those identities, which always felt like an instance of emotional collapse. I refined intricate internal processes for situations when a stray comment or a sideways glance made me feel like my own chamber was being opened and I had no choice but to discard parts of myself in favor of a single state. I wanted to reduce every such instance to no more than a dim peek.

When I was a teenager, I was pained by my state of superposition between belonging to the home that I was born into and the place that I moved to, which was quickly becoming home as well. And I was angry about it. Adulthood dulled that anger; I became too busy with the mundanities and minutiae of building a life to maintain it. It takes energy to keep rage boiling at the pit of your stomach. I turned it into a quiet ache, a sense of impending tears behind my eyes, a thudding in my temples, a heaviness in my bones. I kept telling myself that I should be grateful for having the kind of life that moments of collapse could disrupt at all. This became a mantra I kept repeating. I stopped going home every year. In recent years, had you run into me on a work trip, I'd have talked to you as if New York were my only "back home."

In college, there is a day where, for the first time, I wake up and my immediate thoughts are in English instead of Croatian. This is the scariest collapse yet. Then, it is the most relieving.

My first green card came in the mail in the spring of 2021, after more than two years of low-simmering anxiety, thousands of dollars spent on legal and document fees, and one awkward interview. I was awarded this document, and the status of a legal permanent resident, because my spouse is an American citizen, but this meant that this green card was conditional. After just two years, I would have to re-submit the paperwork and wait for my legal status to be reassessed. Despite knowing this, I was still tremendously relieved. By this point, I had spent a whole year living in New York, had moved in with my spouse after years of maintaining our relationship long distance, was settling into a job as a high school teacher, and was beginning to see my freelance writing career take off. The green card had been the last thing that I needed in order to declare my situation stable. My American life felt like it finally had a shot at being sustainable. Standing at the front of the classroom the next day, I felt a little more grounded in my role as a person with authority, and a person who would get to keep that authority for a while.

My students knew that I was foreign because I told them about it at the beginning of each semester as part of an effort to build trust. But part of me always hoped that they would overlook that detail, maybe even fully forget about it in favor of remembering more of the physics that I was teaching. To be effective as an educator, I thought that I had to avoid the instability of the catlike collapse as much as possible.

But teaching, by default, means always being observed, and hence always in danger of collapsing. During the last few weeks of the school year, a classroom discussion about how a person's identity as a physicist intersects with their other identities—an activity that I had prepared for Pride Month—opened the door for several students to share what being an immigrant, or from an immigrant family, had been like for them. I braced myself. Then, in words far simpler than my yearslong efforts to over-intellectualize myself, students described feeling like they were one person with their parents and relatives, and another at our school. The transition between the two, often a literal bus or train ride from a distant part of the city, felt jarring to many.

Remembering that I was supposed to be the adult in the room, I calmly shared that I often felt that way too. Surprisingly, instead of that sinking, deflating feeling of collapse that I had feared, my chest filled with solidarity. Then came a feeling that I hadn't experienced since I'd first realized I was *gemischt*: the feeling of righteous anger.

I realized that trying to evade or control my tendency to collapse had made me lose sight of an important part of the cat metaphor. Schrödinger's thought experiment requires an experimenter—an outside force that constructs the chamber and the diabolical device. And it is the experimenter that demands that the chamber eventually be opened. The cat is just a cat until it is forced to become a symbol of something unsavory and burlesque. The same was true of me and my students—the state in which we had found ourselves resulted from living in a world marked by harsh boundaries, us-against-them logics, and little tolerance for complexity and in-between-ness. As I write this, in 2025, increasingly, showing any sign of collapse into un-Americanness opens people to the danger of legal prosecution or even forceful detention or removal from the country, as raids by Immigration and Customs Enforcement agents have begun to target US citizens based on faulty perceptions of foreignness and thinly veiled racial profiling.[30]

Reading about these raids, or about a youth baseball team from Venezuela that was denied travel visas for attending a tournament in the US, or about beloved community members across the US who are now being deported decades after arriving here, I cannot help but think about my former students and how this environment in which they are meant to be reaching adulthood may end up shaping them. How much fear and anxiety will become associated with their own ability to maintain, or not, a perfectly American-seeming persona? The chambers, the walls, the various devices that keep us apart and sorted—they are all only starting to feel more visceral and more inescapable.

Yet, just as Schrödinger's diabolical device was something made up, so too are the borders of states and the requirements we must fulfill in order to cross them. And just as Schrödinger wanted to establish himself as an authority on the correct understanding of the world, so too do cultural differences, and the intolerances that often get attached to them, serve to lift one type of person to the top and push everyone else to the bottom.

Wavefunction collapse may be a real mystery of nature when it comes to atoms and electrons, and I hope researchers find ways to shed more light on it in my lifetime. But that's where thinking about the cat stops being useful. In real life, what exactly happens to a person smeared out and split between two cultures is much less urgent than who created and enforced the conditions that caused that split to begin with. I could not see this while looking at just myself, but hearing from students whom I loved and felt deeply loyal to made it all click.

This scene may be easier to imagine: I am having a very early breakfast with a friend at a Turkish restaurant. The walls are painted teal, and our booth has brown leather seats. We are drinking coffee from tiny cups illustrated with the evil eye, and we also have beet and carrot juice in plastic cups equipped with paper straws. As two queer immigrants who work in the media, we have a lot to rehash from our respective weeks and a lot to commiserate about. In other words, we joyfully complain to each other. Eventually, our conversation turns to home and to family, and I am suddenly doing what I always do—blathering about physics metaphors to explain my sad immigration feelings. My friend, both hands securely attached to a breakfast sandwich served in a burek instead of a roll, is kindly enduring my carrying on. "Yeah," they say, in between bites. "I always feel like I am two people at once."

Their tone is calm and disarming. They are all mixed up and smeared out, too. We sit in the un-collapsed in-between together for a second. I love us for it.

CHAPTER 6

PATH INTEGRAL

"If I write an integral like this"—he scribbled a ∫ followed by some equations—"you know what that means, right?" asked my fellow teaching assistant, a few years my senior. It was 2016, and his Bieber-esque side-swept hair covered much of his eyes. He was writing down a mathematical formula that I hadn't seen before. We were both spending the summer working at a program for advanced high schoolers, sponsored by our university's math department. I had just told him that I wanted to enroll in a graduate-level quantum field theory class in the fall, and he'd taken it upon himself to determine whether I could handle it.

He had done advanced work in physics and was considered one of the brightest mathematical minds at our very competitive college. When he wrote down the integral, he was effectively giving me an unsolicited test, but I didn't object. I understood almost immediately, however, that I had failed it. This made me feel so out of my depth that, come fall, I didn't enroll in quantum field theory. I settled for a graduate-level quantum mechanics class and learned a lot. When the mysterious integral reentered my life a few years later, during graduate school, the knowledge base that I had built in college, partly by not skipping ahead to quantum field theory too quickly, really helped me. I finally understood the integral. That older student passed away a few years later, and I regret that I never told him how his moment of tough love ended up serving me well.

Integration—the operation of evaluating an integral—is like addition, but instead of using plus signs you write a symbol that looks like an elongated letter *S*, for *sum*: ∫. When fast-talking professors write it on blackboards, or when you write it countless times while calculating through the night, that symbol can look exceptionally defiant for just a squiggle. The integral that defeated me that summer, however, was not just difficult. It had a name and a provenance. It was the "path integral," and its origins are usually traced back to a 1948 paper by the American physicist Richard Feynman: "Space-Time Approach to Non-relativistic Quantum Mechanics."[1]

Often described as "the best mind since Einstein," Feynman's lore among physicists is unequivocally strong—he is beloved as an incredibly creative thinker and a quirky genius with lots of boyish charm and a propensity for mischief. His path integral inherited some of his notoriety.[2] Even before I formally learned quantum mechanics, I knew that if I stuck with physics, I would eventually have to confront this mathematical object. Professors like to mention it too. Capitalizing on Feynman's popularity, they dangle in front of their students the prospect of learning the "Feynman path integral," presenting it as a carrot you may get to sink your teeth into if you stick with your studies and work your way up to advanced courses, such as quantum field theory.

It is then somewhat ironic that in his paper, Feynman admits that the path integral contains "no fundamentally new results" and instead offers the "pleasure in recognizing old things from a new point of view."[3] The famous path integral is not a discovery of something new in the quantum realm—it is more like a new perspective on the behavior of its inhabitants, a vibe shift of sorts. And Feynman did not invent the path integral from scratch, as a few philosophers had previously discussed similar ideas. Paul A. M. Dirac, one of the founders of quantum theory, had also come close to recasting quantum mechanics in terms of the path integral. But Feynman did present the most polished and unified narrative about it.[4]

The path integral hinges on an idea from quantum mechanics that originally made the theory a little shocking and a whole lot new: our knowledge of what a particle does when we are not looking at it, or measuring something about it, is limited. And there are many things that a particle can do in between measurements. Consider a particle that travels from point A to point C. When it comes to saying how exactly it did that—which exact path it took—the best we can do is to calculate some probabilities. We can say that there is a small chance that it traveled in a succession of loop the loops and a large chance that it traveled in a straight line, but unless we measure its path with an instrument, we cannot be sure.

In his 1948 paper, Feynman considers measuring the position of a particle that is moving along some path three successive times, or making measurements called A, B, and C, each of which records some coordinates in space. He shows that if you assume that measurement B, which happens after A and before C, yields a definite position (marked by concrete figures), that supposition determines how you calculate the outcome of measurement C, or where the particle may end up. Feynman writes that the possible outcome of measurement B matters even if you chose to not actually perform it. He follows that with an even more counterintuitive statement: "We are led to say that the statement, 'B had some value,' may be meaningless whenever we make no attempt to measure B."[5] This is a notion that haunts much of quantum theory: we cannot always claim to know that objects have definite properties, such as having a single position in space, unless we measure them.

So, it matters where the particle is when you take out your ruler or some other measuring device and set out to fill in figures under "position B" in your lab notebook, but it also might not be a settled reality until you actually touch the particle. Yet Feynman, who is often quoted quipping about how no one understands quantum theory, does not dwell on this.[6] Instead, he accepts it and plays along, ultimately positing that if we want to know how the particle can make it from a position determined by A to one yielded by measurement C, we should combine and tally all the possible paths it could take between the two. Whatever could happen when you set out to make measurement B must be part of the consideration, and all the jagged, disorienting, loopy, unnecessarily roundabout paths become part of the calculation alongside the more straightforward ones. They all get summed together by the integral. In the process, each path gets assigned a weight, similar to how a teacher may calculate a final grade for a course, which guarantees that whatever formula the integral simplifies into after you have calculated it is physically true.

This technique is often used to compute a function called a "propagator," which is a mathematical expression, or a formula, for how a particle will evolve over time—how it will reach its future. All the paths it could possibly take, regardless of whether it actually does take them, become part of the recipe for the particle's evolution. Furthermore, computing the propagator does not only determine the particle's trajectory in space and time, but also allows you to obtain its wavefunction, a mathematical expression that contains information such as the particle's energy or spin, for a future time too.

In an interview from 1998, Feynman's contemporary, the physicist Freeman Dyson, put it like this: "You start out with the world in a certain state and then it does everything it can, it can move in all sorts of directions, all the particles can jiggle around as much as they want in all possible ways, and then at the end then there's a certain probability amplitude for finding a particular state in the future, and you find the probability amplitude for that future outcome by simply adding together all the paths. That's it—that's the recipe."[7]

Dyson himself was initially mystified by this method, but the two men both spent time working at Cornell University in upstate New York, so he had a chance to observe Feynman work and live—and to study his work in depth. In the interview, Dyson's demeanor makes it clear that he and Feynman had different ideas on how to do physics and how to be a physicist. He does not go as far as calling Feynman rebellious, but the implication is there.

Despite Feynman's popularity, there is also lots of historical evidence—including in his autobiographical writing—that he was rather sexist, and that he was violent toward at least one of the three women he married in the course of his life. There are anecdotes about Dick (as he liked to be called) calling women "bitches," expecting sex in return for drinks or sandwiches, and marveling at how the "female mind" can only understand analytic geometry in the context of knitting patterned socks.[8] He has often been presented to me as not only the last truly charismatic figure in physics, but also as a science communication role model. Yet I could never look up to him. It's just too hard to believe that when Dick gave endless lectures on the laws of physics, he did so with someone like me in mind.

But I do find emotional resonance in the story of his path integral. I empathize, rather deeply, with the zigging, zagging, looping, and detouring quantum particle.

I've considered three points in my life at which it would have been informative to measure my coordinates and take stock of my state of being. In 2014, you would have found me in Urbana, Illinois. In 2025, a measurement would have yielded New York City. Simply arguing about an entity's position is reductive—it does not come close to offering the complete picture—but even when I make that severe simplification, I can barely grasp just how many different paths could have taken me from one place to the other.

PATH I

POINT A: 2014, UNIVERSITY OF ILLINOIS URBANA-CHAMPAIGN. (Enroll in a PhD program, put in six years of work, publish half a dozen papers in peer-reviewed journals, present at multiple conferences every year, rack up lots of teaching experience and points for "academic service," date your partner long distance for five years. During your nightly run, repeat this mantra to yourself: *I will make progress on braiding anyons and uncovering multifractality in coupled topological chains and there will be an ultracold atoms experiment on the International Space Station directly informed by my calculations, and people will care.*)

POINT B: 2020, NEW YORK CITY. (Graduate on the eve of the pandemic lockdown, get stuck in New York, finally reunite with your partner but under grim circumstances. Persevere. Elbow your way into a remote postdoctoral position at a local university, stop teaching, publish more, get better at computers, review more papers for academic journals, present more, work more and more and more. As you run down a littered Brooklyn path every morning, repeat this mantra to yourself: *These are the best years of my life because I am able to commit everything to my work.*)

POINT C: 2025, NEW YORK CITY. (Apply to dozens of tenure-track positions and a few non-tenure-track ones, fly to a few interviews out of state; end up at a small college in the city, mostly teaching, mostly writing grants, emailing your old graduate school adviser about work from a decade ago as your one lifeline to research that still feels cutting-edge. Anxiously watch your partner secure a similar position, his Ivy League degree unable to protect him from low wages and endless teaching assignments. Get tenure and start spending more time advocating for underrepresented students and agitating for unionization efforts. When you work out among beefy students in the university gym, repeat this mantra to yourself: *I am grateful to live with my partner and I am grateful to be called "professor."*)

One of the most surprising things about the path integral is that mathematicians can't make sense of it. This is not because they disagree with the story it tells about a particle, or because they have philosophical issues with quantum mechanics, but because as a mathematical object called "integral," the thing that Feynman formulated doesn't actually live up to that name. Were a mathematician presented with a path integral outside the context of physics, they would likely question whether it could be used to perform integration at all.[9]

The most ordinary of integrals computes the area underneath some curve on a graph, the curve representing some function. The operation of integration is equivalent to dividing that area into many thin rectangles, then adding them all up. Some of the rectangles' edges will fall above or below the curve, so their sum will be an approximation of the area rather than an exact figure. But if you make the rectangles infinitesimally thin, their edges correspond perfectly with the curve, and the area you get by summing them becomes exact. Now you can stop saying that you are summing and start saying that you are integrating.

In the case of the Feynman path integral, you are summing all the possible paths that a particle can take, and he did work out how to express all of them in one function. But working out what the equivalent of drawing those rectangles would be, and whether each should matter equally—that is, have the same weight—is an open question. Physicists have developed tricks that work for some path integrals, such as assuming that the whole world fundamentally lies on a grid, and have gotten a lot of use out of them. But very few would dispute that they are indeed tricks instead of surefire ways to make rigorous calculations. In other words, it is not clear that at some future point a mathematician could not prove that these integrals do not actually mean what physicists say they do.

The Clay Mathematics Institute in Cambridge, Massachusetts, maintains a list of unsolved problems called "millennium problems," offering a million-dollar prize to anybody who solves one of them. One of the millennium problems challenges researchers to formulate a rigorous mathematical explanation for the range of masses that particles can possibly have according to a specific theory of quantum fields, a task that involves figuring out how to do some path integrals rigorously, with as much certainty as multiplying two numbers together.[10] "I started out my career deeply in thrall to the idea that the path integral was the correct way to formulate quantum mechanics and quantum field theory," Columbia University physicist Peter Woit wrote on his blog in 2023.[11] He goes on to describe how he spent five decades grappling with the mathematical issues concerning the path integral. Now, he teaches it to students only sparingly, and with lots of caveats. Not all physicists swing so far from jubilation to skepticism when it comes to Feynman's work but it is not necessarily atypical to become less enamored with it over time.

The low-simmering drama surrounding the path integral has always amazed me. My training as a physicist included a lot of instruction on how to make approximations. The mathematical equations that physicists produce as descriptions of the world are often too difficult to solve exactly, so it is not unusual to resort to making simplifying assumptions that also introduce a small amount of inaccuracy. For instance, you might treat a very thin object as perfectly two-dimensional, or approximate a bumpy round shape as perfectly spherical. With the path integral, however, it is simply not clear what assumption you would have to make in order to make it more well-defined.

At the same time, as a tool in the physicist's toolbox, the path integral has been tremendously useful. A lot of what we think we know about our physical reality at the level of its constituent parts—particles and fields—

has been built with the help of the Feynman path integral. The Standard Model of particle physics catalogs all the particles that exist and how they interact with each other. If you look it up, you may first find a table of particles, but a little more reading will lead you to functions called "action" or "the Lagrangian," both of which get plugged into the Feynman path integral.[12]

Some physicists believe that the path integral could also be necessary for combining quantum theory and Einstein's theory of general relativity. The two theories are exceptionally accurate at describing different parts of our universe—general relativity works well for cosmically large objects and quantum theory governs the atomically small—but physicists don't fully understand what happens when the two meet. Here, the path integral could be used to sum up every possible shape of spacetime. Each shape dictates what gravity can be like in that version of the universe, so summing them all up would be roughly like combining many possible gravities in the hope that a true one emerges. If this approach worked and allowed physicists to derive gravity in a quantum way, the feat would be nothing short of revolutionary. Stephen Hawking was a proponent of this approach, and researchers are currently pursuing it.[13] In so many ways, the language we use to describe the universe is dominated by the sloppy dialect of the Feynman path integral.

In an objective sense, it is meaningless to use quantum theory to talk about the path of a person. We are too large and too warm for quantum phenomena to affect us. Unlike a particle, we also must reckon with having free will and being able to choose our paths ourselves. But for whatever reasons, we do sometimes take strange routes between different points in space and time. And their strangeness does not always mean that the simpler, more straightforward trajectories would have been more correct.

Assuming that I am currently at some semblance of point C, I often wonder what the person I was at point A would have to say about me and the path that has taken us here. In the last ten years I have worked as a teacher, a researcher, a project manager for a theater production, a labor organizer, a community advocate, and a journalist. I was a girlfriend a few times, a somewhat secret lover once, and eventually a wife, though now I prefer to be called a spouse. I wore all black and listened to lots of heavy-and-heavier metal until I started buying Lucy Dacus records on vinyl and wearing pastel slacks and floral button-downs. At one time, I tried to look girly and feminine, with knee socks and short skirts and corsets and red lipstick, then realized I was not a woman at all. I made fun of gym enthusiasts until I was myself running at least five miles every day, shoulder- and chest-pressing small dumbbells in my kitchen a few times a week, and paying for weekly classes in boxing and hot yoga. It is hard to wrap my head around how disparate these points of my identity have been, yet I lived them—and it must have been me who moved along the paths that tethered them together. And then there were the many paths that I did not get to take, or did not want to take.

If you were to tally up all these paths—try to let them cancel or amplify each other in the spirit of the Feynman path integral—what would you end up with? I imagine a room filling up with different versions of me until they all start to shout and compete with one another. At that point, I have nothing left to do other than cover my ears and hope that a bigger truth will eventually emerge from that noise.

PATH 2

POINT A: 2014, UNIVERSITY OF ILLINOIS URBANA-CHAMPAIGN. (Enroll in a PhD program, put in six years of work, publish half a dozen papers in peer-reviewed journals, present at multiple conferences every year, rack up lots of teaching experience and points for "academic service," date your partner long distance for five years. During your nightly run, repeat this mantra to yourself: *I will make progress on braiding anyons and uncovering multifractality in coupled topological chains and there will be an ultracold atoms experiment on the International Space Station directly informed by my calculations, and people will care.*)

POINT B: 2020, NEW YORK CITY. (Graduate on the eve of the pandemic lockdown, get stuck in New York with your partner but without a job, teach at your old school remotely while furiously applying for jobs, pick up some freelance writing work because you finally have time, persevere, score a teaching position at a high school early-college program, accept the job just as schools start reopening. As you run down a littered Brooklyn path every morning, repeat this mantra to yourself: *This is the most impactful and ethical work I could ever do. If I cannot do physics, I ought to teach it well.*)

POINT C: 2025, NEW YORK CITY. (Keep teaching, first ninth grade and some typical college classes, then more and more electives. Struggle your way through a master's degree in education, earn tenure with New York's Department of Education, love your students no matter what, give them the best instruction possible by trying new things in your classroom even when your supervisors care more about school politics than the well-being of your students and your classroom is under-resourced. Keep writing, but just a little. Coach the cross-country team. Stop dreaming of a research career, stop dreaming of a freelance writing career, write a textbook, become someone whom younger teachers can come to for advice, plan to retire before you hit seventy. As you run behind your students in a park transformed by the city's efforts to make it resistant to climate change, repeat this mantra to yourself: *I have made a difference here.*)

A room teeming with different versions of me has long been a useful visual for my conflicting feelings on how I've moved through spacetime throughout the years. For a while, I tested the correctness of my current path by trying to guess how my sixteen-year-old self would judge it. They were the person who'd found the courage to move us across the world, and that act came to signify a self-confidence that I felt had slipped away from me in later years.

For a while I could imagine earning that person's approval—during my first year in graduate school, for instance, when I learned how to calculate path integrals and started the research projects that would later define me as an academic. The sixteen-year-old had wanted us to attend a different university for graduate school, but I was still checking off a lot of their boxes by being in a PhD program at all. But time marched relentlessly forward and mercilessly brought change. The older I got, the more I had to make excuses for having fallen short of that young person's expectations. The versions of me who'd taken a different path, who'd hit different points between B and C, would have less to apologize for, I thought. I grieved not being that person.

A lot of feelings factor into these internal confrontations with the versions of me that zigged, zagged, and looped differently. Some of it is shame at having failed to accomplish some very specific thing. Some of it is the worry that I have wasted time traveling many twists and turns along a path that was always going to take me where I am now. I could have saved myself some trouble had I known I'd end up here. I could have lived a better life had I known earlier who I now know myself to be. Some of it is just the heaviness of imagining someone who is almost me, who dislikes me, who disapproves of who I am.

In 2020, a few months into pandemic lockdown, I was applying to many academic jobs but felt qualified for none. I felt like the path in which I had invested all my life up to that point—the path that was supposed to end with me becoming a professor at a research institution—had disintegrated. This was always a likely outcome—academic jobs, and especially those designed to end in tenure, are scarce. The number of people who obtain a doctorate in physics is greater than the number of those who use it to become tenure-track professors by roughly a factor of five.[14]

I was also seeking jobs in a field that had inherited its culture and structure from people who were nothing like me, so any success I may have had would have always been in spite of the system and not because there was a place for me within it. A study published in *Physical Review Physics Education Research* in 2022 found that, in a survey of LGBTQ+ physicists, 36 percent "reported considering leaving their institution or workplace."[15] I may have been in that group in the last few years of my doctoral training, but after I graduated it became apparent that just *considering* leaving was the better-case scenario. I went through with that consideration and actually left.

It certainly didn't help that I'd graduated just as the pandemic was making job markets shrink and freeze. It was widely reported that, among academics, the pandemic most adversely affected female scholars, and subsequent studies and meta-studies found that the pandemic correlated with an increased gender gap in academic productivity.[16] I am not a woman, but I did do a lot more housework and care work during the most severe stretches of the pandemic. Living in lockdown with my family did pull me away from my academic work. One survey of Canadian academics in 2022 found that "women and non-binary or genderfluid respondents reported higher stress and pessimism than men."[17]

I felt that pessimism in the summer of 2020, as old research projects dragged on and drafts of papers that would only be published after I switched careers sat coffee-stained on the dinner table. Whether I was working on them or finishing cover letters, one thing remained constant—the judging eyes of all those past and alternative versions of myself.

Certainly, my career choices were not the only thing these imagined persons could hold against me, but becoming a professor felt like the one thing we had all aspired to. Giving up on it felt more damning than cutting my hair or my changing taste in music. The other Karmelas could have forgiven me an awful lot had I just stayed on the straight and narrow professionally. In my mind, the job I wanted to have was so severely merged with who I wanted to be, and who I thought I was, that the prospect of having to do something else felt like a devastating failure by default. I later learned that there is a term for this: I had become "enmeshed" with my career.

"Psychologists use the term 'enmeshment' to describe a situation where the boundaries between people become blurred, and individual identities lose importance. Enmeshment prevents the development of a stable, independent sense of self," wrote psychologist Janna Koretz in the *Harvard Business Review* in 2019.[18] She goes on to clarify that people don't get enmeshed only with other individuals, but also with their careers. In 2021, psychologist Anne Wilson told the BBC that if you tie your identity to your career, "the successes and failures you experience will directly affect your self-worth," and that "because we live in a society where careers are less likely to be lifelong, if we switch or find ourselves out of a job, it can also become an identity crisis."[19] Koretz identifies several factors that contribute to someone becoming enmeshed: a work culture that encourages long hours and going "above and beyond," a person's family and community expressing pride in their job, and the fact that in some professions, being overly invested in your work can lead to higher income and propel you into a higher socioeconomic class.

Academic careers are typically not lucrative—in fact, graduate and postdoctoral positions often bring a lot of financial instability into a person's life. This occurs because graduate student pay tends to be rather low—even at an institution that recognized a graduate employee union and engaged in collective bargaining, my wages always put me only a few thousand dollars above the poverty line—and postdoctoral positions, which do pay better, rarely last longer than three years and often require several costly cross-country moves. But the factors concerning culture and status that Koretz names certainly applied to me. After all, I had moved to the US to pursue a better physics education and become a physicist, my own nerdy American Dream. This may have doomed all the American-bound versions of me to experience enmeshment at some later point.

The history of physics is lousy with great men who hailed from the upper—even aristocratic—classes of society, or who moved into them by getting an education. The people who uncovered the inner workings of the universe by studying stars and planets had to have money to build telescopes and leisure time to look through them. Some were men of religion and members of the clergy. But many of the scientific revolutions of the Enlightenment were also financed through colonialist exploits.[20] The culture of physics today still shows traces of the notion that being a physicist makes you "better than," both because you have elite knowledge of physical reality and because you are connected to powerful people.

Physicist Chanda Prescod-Weinstein writes, "Astronomers were once being funded to watch an eclipse from Haiti in order to better measure distances, ensuring that slaves and goods would move faster across the Atlantic. Today they are funded to help develop adaptive optics, a military technology that makes our beautiful pictures clearer, whether they are spy images or galaxy images."[21] My own experience of working in condensed matter and atomic, molecular, and optical physics is that military funding is ever-present. It is often the lifeline that keeps radical experimental studies going.

The public image of physicists, however, emphasizes characteristics other than their class status and other privileges: the drive to be almost monastic in their devotion to their research, and a very rigid, masculine understanding of what it means to be objective and rational. I was an impressionable kid, and when I was reading about the canonical heroes of physics, I would get too wrapped up in those stories to ask whether they were the only ones out there. I internalized every narrative about the rewards of punishingly hard work and about never, ever bringing your feelings or your identity into your profession.

I bought into the idea that science is done mostly through discipline and devotion, that who you were, where you came from, or how wealthy and well-connected your parents were simply didn't matter. I aspired to what was sold to me as a life fueled by curiosity, rich with hard-earned insights. And having so often been put in my place by adults who wanted me to stop being so loud, so opinionated and conceited, I also wanted to gain scientific wisdom that they would have to respect. Even at sixteen I was not immune to cravings for prestige and power. I never fully satiated them, but I definitely got enmeshed.

Enmeshment is not unique to academic science. More often it is discussed as an affliction of lawyers and entrepreneurs. This points to a more unsavory truth: within capitalism, workers are more easily exploited if their identity is inseparable from their job. In the modern era, many people have some choice over what they do, and convincing them that they can be happy if they just find their calling has become an industry in itself. At the same time, research finds that young people are negatively affected by the pressure to find the profession that will express their one true purpose.[22] A quest to discover your dream job is compelling because it promises to end in a life where all your days are ripe with fulfillment. In reality, it conflates people's personal identities with their roles as workers, which makes them more miserable while bringing profits to their bosses more seamlessly.

I didn't fully understand this until I left academia's ivory tower but was still surrounded by people who claimed that they'd found "their people" on the job, whether that job was teaching or working at a magazine. They would assure me that I was just the right type of person to work alongside them, even when I didn't ask for their appraisal. Conspicuously, being the right type of person always meant a willingness to work longer hours, take shorter lunch breaks, take on unpaid projects, and abstain from picking fights with leaders and administrators. Even if my managers did express concern over how much overtime I worked, they would frame it as a facet of my personality, as if I had become that way all on my own, with no influence from our shared work environment. I still have these conversations. If I had been born a decade earlier, someone would probably be affectionately calling me a "workaholic"—a word that has since gotten a bad enough rap for managers to know to use a softer synonym.

Of course, there is nothing wrong with liking a job, and I do unabashedly love my work and believe that it is worthwhile. But from a mental health perspective, what can go unacknowledged is the fact that all the enmeshment factors that Koretz identifies—cultural expectations, socioeconomic status, approval from family—are hardcoded into the political and economic system that we live in. My desire to be a scientist was not immune from the ills of capitalism, because the only model of becoming a scientist available to me was one that was shaped by capitalism. The jobs I've held outside of science followed the same pattern.

And I am certainly not the first person to realize this. Nearly two hundred years ago, Karl Marx, in his "Economic and Philosophic Manuscripts of 1844," argued that workers could only find fulfillment in their work if they could control their own labor and the products of it.[23] What millennials like me have been given instead is an illusion of that fulfillment—an illusion that we do have control of our labor and just choose to perform utterly too much of it.

When I was in graduate school, a faculty member once told me that their postdoctoral years, when they were employed as a researcher with no administrative and teaching responsibilities, had been the best time of their life. They'd worked ten or twelve hours a day, they said, but had still gotten to hike on weekends. It had been great to do so much physics and to be so close to nature, they reminisced. I wanted to respond by saying something about nature, but their comment about ten-hour days rendered me speechless. And I felt like a failure for having that reaction.

"But don't you love physics more than anything?" whispered some alternative, more resilient version of me.

In the next few years, I would go on fewer than half a dozen hikes, but I can assure you that my count of ten-hour days was much, much higher.

PATH 3

POINT A: 2014, UNIVERSITY OF ILLINOIS URBANA-CHAMPAIGN. (Enroll in a PhD program, put in six years of work, publish half a dozen papers in peer-reviewed journals, present at multiple conferences every year, rack up lots of teaching experience and points for "academic service," date your partner long distance for five years. During your nightly run, repeat this mantra to yourself: *I will make progress on braiding anyons and uncovering multifractality in coupled topological chains and there will be an ultracold atoms experiment on the International Space Station directly informed by my calculations, and people will care.*)

POINT B: 2020, NEW YORK CITY. (Graduate on the eve of the pandemic lockdown, get stuck in New York with your partner but without a job, teach at your old school remotely while furiously applying for jobs, pick up some freelance writing work because you finally have time, persevere, score a teaching position at a high school early-college program, accept the job just as schools start reopening. As you run down a littered Brooklyn path every morning, repeat this mantra to yourself: *This is the most impactful and ethical work I could ever do. If I cannot do physics, I ought to teach it well.*)

POINT C: 2025, NEW YORK CITY. (Keep teaching, first the ninth grade and some typical college classes, then more and more electives, struggle your way through a master's degree in education, earn tenure with New York's Department of Education, love your students no matter what, give them the best instruction possible by trying new things in your classroom even when your supervisors care more about school politics than the well-being of your students and your classroom is under-resourced. Keep writing a little, first on your lunch breaks and at night, then more, on weekends. When editors start reaching out to you, never say no, join groups for science writers, pitch more bravely, post links to your articles more often, and when an editor messages you looking to hire a physics writer for a brand-new newsroom in New York, answer them. Stay cool through two rounds of interviews, tearfully quit teaching, become a full-time physics journalist. As you run on a treadmill in a grimy neighborhood gym in Queens every night, repeat this mantra to yourself: *Even if science journalism crumbles tomorrow, I will be able to reroute again.*)

Of all the cover letters that I wrote during that panicked job hunt in 2020, my favorite was one that I privately referred to as "Whoever you need me to be." I sent it in response to a few posts for jobs that were listed under the vague heading of "community manager." A career counselor at my graduate alma mater suggested that hiring managers might overlook my painfully homogeneous academic job history if I sounded earnest and courageous in my cover letters. "Use your writing skills to show off your personality," they said, and I took their advice to heart. Here is what I wrote:

> Early in the week, I watch a paperclip levitate inside an MRI together with an artist who asks me what it might feel like when the spins of all the atoms in your body simultaneously flip inside such a machine. This discussion will factor into a performance piece I am production manager for, one that features actors, physics students and a cameo from a Nobel laureate. The next day, I attend a meeting of the Graduate Employee Organization and present a part of the budget I have worked on with the Finance Committee. We push for a clear membership number goal and plan a big drive to sign those new members up. A few weeks later, I will be one of the people knocking on office doors and engaging many of them one-on-one. Midweek, I teach introductory quantum mechanics to future engineers and they ask me how a bunch of differential equations relate to semiconductor technology that makes smartphones work. In the afternoons I dive into other equations for my own research and daydream about ultracold atoms in space. At the end of the week I call into a meeting for a conference I am co-organizing with students and

researchers from eight other universities interested in mentoring and diversity advocacy in science. First, we discuss the best way to order Thai food for over fifty attendees while keeping everyone safe and our spending reasonable. Next, we break into groups and a younger organizer discusses an Impostor Syndrome workshop we are co-developing for the conference. On Friday night, I practice yoga, make bread that needs to sit in the fridge overnight, then work on an essay for a newsletter I will be sending out that week.

Did all of this actually happen during one seven-day block? Maybe not, but it is certainly what every one of my weeks has felt like for the past six years.

Though I have always wanted to be a scientist, working on my Ph.D. has turned me into a person whose head fits one more hat every year. I became a teacher and an educator, then a mentor, then an advocate, then an organizer and finally a science communicator and a non-fiction writer. I have given talks about condensed matter physics and on how to plan team building overnight retreats. I kept a lot of heavily color-coded spreadsheets and practiced explaining the Pauli exclusion principle without ever saying the words "field theory." I mentored artists in developing dances based on black hole mergers, coached physics students in writing poetry, and once carried two very heavy chairs across campus because they were a stage prop that a physicist-turned-performer really needed. I always considered myself to be a quick learner and good with managing my time, but the complex web of tasks that was graduate school—ranging from communicating science through my teaching and outreach work to managing complex projects like organizing a conference and working on diverse teams of researchers—pushed me on both fronts and prepared me for a variety of future challenges. I would relish the opportunity to apply the skills I gained as a project manager, a community builder and a lover of science by being part of your team.

I was not offered any positions based on this letter, or even any interviews. One rejection email did come with a sweet note about my writing and my passion, and I took that to heart. I had not grown up wanting to be a community manager and probably would have quickly burned out in such a role, but this cover letter is my favorite because it not only reconciles all my different paths but also frames their clash with one another as a generative process.

There are many ways in which I have moved and could have moved from age sixteen to where I am now, and they all matter. Some are worth remembering as missed opportunities and mistakes, others as informative hypotheticals. Some were briefly my reality, others may only become real as I move deeper into my future. I am writing this halfway into my thirty-third year, and this version of me reflects that totality.

Enacting what psychologists call "the end of history illusion"—one's belief that they were very different in the past, but will not change much in the future—I imagine that the present is my point C.[24] My time coordinate is the year 2025, my space coordinate is New York, and the rest of my parameters are set by being queer and being a journalist, and by the myriad things I do every day. And I believe that the current version of me is here to stay—that all those other versions of me shouting at each other are about to finally shut up. But it's called an illusion for a reason. A point D almost certainly looms large on the horizon. The long reckoning between who I was, who I could have been, who I thought I would be, and who I find myself being could resume at any time with just a slightly different cast in that room where all the voices are mine.

But by calling it a reckoning or a confrontation, I impute violence to my past, hypothetical, and present selves. I suggest that one of us must be correct and that one of us must win. This too is a function of having grown up in the winner-take-all culture of late-stage capitalism, with a sense of having to outwork and outdo everybody else to gain approval, status, and self-fulfillment. I don't want to care about that kind of winning anymore.

To find a better way to resolve this psychological path integral of mine, let's look at the double-slit experiment, first performed in 1801.[25] Here, light is shined through a barrier punctured by a pair of two very narrow adjacent slits and onto a screen behind them. One would expect this to create two bright spots on the screen, one behind each slit, but the actual result is much richer. The screen fills up with alternating bright and dark stripes, each pair separated by many shades of gray. This is because light is a wave, so instead of traveling through a slit and onto the screen in a single point like a bullet, it spills through the slit and interferes with the light wave that has similarly spilled through the neighboring slit.

Imagine each wave as a series of peaks and valleys, where the higher peaks are the brighter light. When a peak of one wave meets a valley of the other, they cancel out and you end up with darkness. A peak meeting a peak, on the other hand, produces very bright light. The in-between values of brightness—the shades of gray on the screen—are created by similar overlaps: the edge of one peak and the edge of one valley, for instance, or two slightly misaligned peaks. All quantum mechanical objects can demonstrate this sort of wavelike interference. The process of calculating the path integral, then, is like adding more and more slits between the two you started with. For the object, each new slit creates a new path to the screen, adding to the interference effect. Eventually, you'll cut enough slits—maybe millions—for the barrier to cease to exist. The space it once occupied will now be filled with the object traveling along every possible path.[26] The formula to which an object's path integral reduces will capture the image that the equivalent of a screen would record here.

The striped screen in the double-slit experiment offers me a metaphor for the clash of selves. A clash or a conflict does not feel like something that could produce signals in the in-between spaces or turn out anything that reads as a shade of gray. What feels more appropriate is to imagine a negotiation, a communal effort where valleys of one Karmela can be ameliorated by the peaks of another. In solidarity with all the other versions of myself, I may stand a chance of wresting my values away from the exploitative system that I live in, and I may even find courage to challenge it. If all the interfering Karmelas seek connection rather than mutual destruction, maybe their final product—me as I am now—can be imbued with generosity and openness as well.

The path integral owes a lot to Feynman's predecessors, but also to his famous intuition—and his belief in that intuition. The combined intuitions of every me that ever could exist must count for something too.

CHAPTER 7

RIGID BODY DYNAMICS

CLASS ASSIGNMENT 1

Conceptual Physics II
Spring 2021
Dr. Padavic-Callaghan

In this project, you will use ideas from physics to explain what happens in some sporting event.

For example:

When you kick a soccer ball, you impart a force on it over some time—you are applying some impulse. This changes the momentum of the ball and can also change its velocity and acceleration. How far the ball flies after you kick it depends not only on the strength of the kick, but also the angle, since momentum, velocity and force are vectors.

If the ball hits another player, some of its momentum will be transferred to that player and they will also have to apply some impulse in order to change the direction of the ball or make it go faster.

One example is when a goalie catches a ball—some of the ball's momentum is transferred to the goalie, and they fly backward a little. The goalie moves slower than the ball because they have a larger mass. Here, conservation of momentum helps us see that the velocity of the goalie after the catch will be smaller than the velocity of the ball before the catch.

KNOTS

In 1867, the physicist William Thomson suggested that the world was made of "knotted or knitted vortex atoms."[1] At the time, physicists believed that all of space was filled with an invisible substance called *aether*, and Thomson proposed that his vortices—tiny, elongated tornadoes and whirlpools—existed within it. He was wrong on two counts: by the 1890s, physicists discovered that what's inside atoms is not knotted whirlpools but particles like electrons, and in the decade that followed, physics would also fully discredit the idea of ether. But knot theory, the mathematical study of knots, lived on. Now it even has practical uses, such as using its equations to work out the best way for an aircraft to switch orbits, and, in cryptography, helping to verify the entity that encrypted a piece of information.[2]

I learned a little about knots in college when I was a teaching assistant at a summer math program for high school students. The basics seemed very simple, but after a few weeks the complexity of the problems increased sharply, and I had to do a lot of studying of my own before stepping into the classroom. But the kids had fun, and I enjoyed the challenge.

The summer before, or maybe the one before that, instead of learning about knots, I learned about my heart. My heart, I was taught, was marked by a small hole. This happened in the living room of an energy healer in a small Croatian town—a healer who, according to my mother, everyone on our island wanted to see. My mother had brought me along to an appointment she'd made for herself, probably hoping that the healer would work on me too. And he did. His diagnosis was ultimately far more complex than simply that I have a tiny hole in my heart. (Later, my mom admitted that a doctor did tell her about this hole once, but it seemed so benign and insignificant that she forgot all about it until the healer brought it up.) As his hands hovered an inch above my skin, he proclaimed that he had found a "knot of emotions" deep inside of me—and that if I did not learn to disentangle it, my future would be filled with darkness.[3]

In the moment, I scoffed at his assertion, but the image of my body filled with knots has since often resurfaced in my mind. Over the years, my body became much more of a site for conflict with myself, a source of discontent, a resting space for pain and self-doubt. It also became much more of an optimization project, by which I mean a project in whittling myself down. Years later, to course-correct all this, I made room within my worldview for the healer's belief in his own esoteric gifts and the spiritual energy he believed resides in the body. This is slightly ironic, as many of my friends and acquaintances know to say that "the body keeps the score" because of the work of a scientist like me, and not some mystic.

Much of knot theory consists of working out how to untangle a particular knot. It is a difficult problem. In the mathematical definition of a knot, the theoretical string from which it is made is always connected so that when you untangle it, you never end up with two separate ends. Consequently, the simplest knot that you can make is a circle, which is often called the "unknot." A knot becomes equivalent to the unknot if you can disentangle it into a loop without cutting the string. For instance, the trefoil knot, which looks like the common overhand knot but with the typically loose ends connected to each other, is the simplest knot that is not equivalent to the unknot. If the unknot is analogous to "0" in the mathematical knot hierarchy, then the trefoil knot is analogous to "1." How many knots are equivalent to each other, and how many are distinct, is an open question in mathematics. The "unknot recognition problem" specifically comes with layers of difficulty. It is not only hard to solve—a knot with up to nineteen crossings, for instance, produces about 21 billion possibly distinct knots that you then must sort through—but it is also hard to predict how much time it may take to solve on a computer, or whether it would be possible to check its solutions if you did manage to get them.[4]

Over the years, I started to think of the thing the healer had identified within me not as a single tangle but as a collection of knots, each with its own provenance. The trefoil knot, born from the pain of a childhood punishment. The star-like cinquefoil knot, made from my fear of being imperfect. The shame of my first boyfriend never introducing me to his friends, neatly coiled in the center of a three-twist knot. The sting of leaving my academic career, tangled into the coils and star-like arms of a stevedore knot. Quietly, to myself, I name them all and assume that they are not unknots, that I will have to find a way to cut and break them someday.

The healer did not provide me with any personal knot theory. Instead, as if we were in the first half of a coming-of-age movie, he suggested that I go to a deserted place of some sort, maybe a forest, and scream the feelings out. I could expel the knots from my body if I just let go of all my stiffness and decorum, he seemed to suggest. That would be the ultimate cutting implement.

In 2023, I interviewed a mathematician who was studying knots of worms. Groups of California blackworms can form large, blob-like tangles to protect themselves from danger, and they can unknot themselves in mere milliseconds. This is so fast that mathematicians who study knots, including the one I spoke to, were stumped by the animals' unknotting prowess.[5] Candidly, he told me that he would have thought this unknotting was impossible had he not seen it himself.[6]

In their studies, he and his colleagues had figured out that the worms were performing a key motion, a sort of a full-body helical wriggle, that helped them disentangle quickly. Talking with him reminded me of another set of moves, one that I'd learned about during that college summer: the three "Reidemeister moves" that you can perform on a knot to turn it into another knot without cutting. Two knots that can be transformed into each other like this are also said to be equivalent. For instance, a knot that looks like a rectangle with two loops at its center can be unraveled through Reidemeister moves into the trefoil so the two are equivalent. This set of moves allows you to twist or untwist a knot's string, to move one piece of string over another or uncross them, all without cutting. Because the worm blob was not made of loops but rather animals with distinct ends, it did not make sense to ask whether the worms were performing Reidemeister moves. But when my interview with the mathematician ended, I sat at my messy office desk and thought about the moves and wriggles that I had learned in order to undo my own tangles.

I never yelled at trees or at the sea or in the shower, or even into a pillow. If anything, over the years I got quieter. Screaming was not going to make my knots budge. But the thing that ultimately did was still a kind of letting go, and a sort of reprieve for the parts of me that I disliked or held onto with fear.

In high school, before I knew about the knots, I'd taken up running. My mind yielded to my legs and the rhythm of my heart, releasing years' worth of shame over being called clumsy and cerebral. I continued to run through grad school, when I also discovered hot yoga. I sweated so much that it felt uncouth, and I trusted the teachers showing me motion schemes that required getting over myself and my fears of looking silly. Then I did strength training, and finally a boxing class, which engaged something primordial in me, something grunting and masculine and strong, much more unbounded and physical than anything I had ever been before. Maybe it was all a bit like screaming, but with my whole body, unabashedly.

My shoulders erupted with a variety of muscles that met each other in clean lines, my core tightened and hardened, my thighs and calves grew. One summer, at a beach back home, my mom said my legs had become *vretenaste*, or spindly. This remark felt different from all the previous times she had simply called me thin, like a compliment that was not conditional on my being small or dainty. During the same vacation, my stepsister asked, "How many squats do you do to have an ass like that?" Later, looking at photos of us from that time, I could see some new strength in my posture.

When I was running, or doing burpees with the tuck jump at the top, or practicing my jabs and crosses, or settling into the Warrior II yoga pose, I felt the strength of my body, rather than how tangled up inside I was. My knots seemed to relax into strings packed with power. I felt hardwired to do something other than harbor and anticipate the emotional darkness that had been sitting inside me even before that healer called it out.

When mathematicians study knots, they use computers, and no computer is big enough or fast enough to keep trying all possible moves on every possible knot. So, researchers turn to complex calculations to obtain numbers called "invariants" for each knot, then compare them. Two knots that have the same invariant are thought to be equivalent. For instance, the invariant called the Jones polynomial has the value 1 for the unknot and all its equivalents. Mathematicians have searched through dozens of trillions of knots to try and find another whose Jones polynomial takes the same value. They haven't found one yet, so this invariant could be a good classification tool, though countless more knots remain to be examined.[7]

Some researchers have also been using artificial intelligence to develop proofs that could add more general, structural knowledge to the classification of knots, instead of simply going through all of them one by one. This is a technology that has a high chance of producing a radically detrimental impact on the environment in the next several years, thanks to its large energy and water demands, and it has not yet led to physics breakthroughs that would justify its broad use in that field. But problems like the classification of knots seem to be a natural fit for the pattern-recognition prowess of programs we currently call AI, so using them could eventually lead to progress in solving this problem.[8] So far, it has only been able to offer hints rather than a full-fledged solution.[9]

When I am not in motion, when I am not sweating, most of my knots are alive and well. Whether they tighten and become permanent seems to correlate with what is invariant in my own life. The healer's words feel the most correct when I hold myself together too tightly, when I close myself off to others too much, when I assume that I am just hardwired to never relax. The knots feed on these sentiments and coil more vigorously in response. To defy them, I go on runs, let friends take me to yin yoga, grunt my way through boxing classes after work, lift dumbbells in my neighborhood gym even though they are Captain America–themed and the men in there look at me funny. Some days, I'd much rather just curl my whole body into one tight twist, and I am still learning whether there might be times when that is the right thing to do. But most of the time, I do make it out the door and wriggle myself out of the darkness.

POWER

Power is the rate of change of energy. To calculate the approximate power of yourself lifting something heavy, you will need a scale, some measuring tape, and a stopwatch.

First, you need to know how massive your something is—that's what the scale is for. Record the mass of your object in kilograms. (If you're using a dumbbell or a bag of rice, just read off the figure printed on it.)

Then, lift the object as steadily as you can. Calculate how many meters it traveled—that is, the height difference between its lowest and highest points. Write that number down too. Next, do the lift again and time it with the stopwatch. If you can, get a friend to help with this. They should start the stopwatch just as you start lifting and stop it right when you get the object to its highest point.

If you are on planet Earth, the last number that you need for this calculation is called *g*—the acceleration due to gravity—and its value is 9.81 meters per second squared, which are the units of acceleration. Again, power is the rate of change of energy, so, in your case, the amount of power recorded shows how much energy you used to fight Earth's gravity divided by how long you were doing that.

So, multiply the mass of your object by the height you lifted it, then multiply the result by *g*. This is the amount of energy that you used. Divide it by the number of seconds you spent lifting. Like the fifteen-year-olds who once took my conceptual physics class, you now know roughly how many watts of power you exerted during your lift. Compare it to a lightbulb if you'd like, just for reference. Don't get discouraged if the lightbulb beats you.

I exert the most power when I throw a "cross" punch in boxing class. Each class consists of ten rounds: in five of them I punch a bag, while in the other five I use dumbbells and a bench for high-intensity interval training. The lights are always low, and the music is always loud and heavy on pop. Over the music, the trainers yell out instructions and encouragements so corny that, once my heart rate gets sufficiently high, I cannot help but believe them. Toward the end of the forty-five-minute session, a black light comes on. If you choose a workout outfit in any light color or featuring white details, just about when you're getting really tired you start to glow.

In my first class, the black light caught me off guard. I was going through a set of sprawls—flinging myself to the ground, then springing back up—and the rapid up-and-down motions were making my heart pound as loud as the music. Then I lit up, like an android suddenly powering up in some sci-fi production. Power coursed through my taut tendons, tingled in my straining muscles, perhaps even crackled in my bones. At the edges of my mind, I felt a sharp hunger to be unstoppable, unbreakable. When it all ended and the lights came on, the person who met my gaze in the studio mirror was glistening with sweat, their body covered in an unbeatable shine.[10]

My boxing studio is part of a sprawling franchise, and its interior is designed for Instagram-ready workout selfies. Week to week, I am being sold an experience that is part of what Jia Tolentino, in her 2019 essay "Always Be Optimizing," identified as a constant need for seemingly empowered women to better themselves in very physical, mechanical ways.[11]

Not everyone in my boxing classes is a woman, but women are in the majority. Locker-room chatter reveals many to be workers in finance, tech, and consulting, many of them relatively young and already quite slim. The exercise classes that are marketed to women are a part of a bigger machine meant to convince them that they can gain power in society through physical self-betterment. It is a two-pronged strategy for preventing women from focusing on collective action or building solidarity with each other and demanding systemic change: becoming physically fit takes time and it makes the focus of all action narrow and individualized. As trainers often say, even in a group workout class it is "you versus you" in the end. In my boxing class, fellow students and I are also encouraged to "show up for ourselves" in the form of "giving 110 percent."

I've boxed on very few hours of sleep, I've boxed to stop myself from quitting a job, I've boxed instead of grieving a broken heart, I've boxed with only half a granola bar and a small yogurt to get me through the afternoon, and I've boxed when I was so depressed that I could barely leave the house without crying. What drove me in those instances was our culture's belief that trying super hard and expunging all personal weakness will save you not just from yourself but also powers much bigger than yourself. As Tolentino writes, "The beauty ideal asks you to understand your physical body as a source of potential and control. It provides a tangible way to exert power."[12] Yet, even if I became the best student in my boxing class, I would still have to worry about paying rent or being disrespected by coworkers. The empowerment that I pay for in that dark, loud room does not help me change my world—it just grants me a temporary reprieve and a false sense of security about how much exhaustion I can endure. Tolentino again, on her experience of barre, which is her preferred workout: "What it's really good at is getting you in shape for a hyper-accelerated capitalist life. It prepares you less for a half marathon than for a twelve-hour workday, or a week alone with a kid and no childcare, or an evening commute on an underfunded train."[13]

Arguments like Tolentino's shed light on the woman who thinks that she can overcome her marginalized status in a patriarchal society by becoming something like a sleek, powerful fembot that updates and optimizes more often than an iPhone, be it through acquiring new gear or working on her body. I have certainly been that woman. That woman and I probably still share many affinities. I have no illusions about the fact that both she and I buy the same product when we sign up for a workout class or a gym membership. Neither of us is exactly buying a revolution. The revolution will not be available through ClassPass. I am aware of this. Sometimes I am ashamed of it too.

But I keep paying for boxing classes, and I keep paying for my gym access. There is something that keeps me in those spaces that is more intrinsically meaningful than the achievement-chasing, optimization-obsessed mores of late-stage capitalism. As much as workout culture catered to women draws from an inherently white and thin beauty standard, boxing also helps me recontextualize my body, which was socialized as female. Even if I am paying for a sense of strength that actually reinforces the system that wants to keep me relatively powerless, I still find my own body to be a fertile ground for subversion.

I was an avid runner way before I found the language of nonbinary gender to describe myself, and I liked how running changed my body. It made me lean and wiry, and people complimented my legs more. But running has also historically been recommended to women who are looking to burn calories without putting on any un-feminine bulk. My experience with running has been tainted with this too, and at one point I was channeling some of the uncertainty I felt about my gender into a desire for thinness. I wanted to turn my body into that of a perfect, healthy, Instagram-able woman, at least in part to convince myself that I could be that—both conventionally perfect and a woman. Running every day helped with that, especially when coupled with restricting my calories. These days, I eat more and worry less, so when I run, as I do most days, the rush of adrenaline at the end of a good five miles feels more purely joyous. But it is still difficult to step on the treadmill or the road without some emotional baggage in tow.

In contrast, by the time I came to boxing and strength training, I was less interested in defeating my body and its "unproductive" need for food and rest, and I was also more confident in my queerness. Hitting bags, lifting dumbbells, doing jump squats and renegade rows—all of this quickly started to feel like a way to free my body from the heteronormative, cisgendered standards of womanhood. I felt less of a need to be perfect and small. Certainly, there are cisgendered and heterosexual women who are incredible boxers, and I am not boxing in order to fully erase all traces of womanhood from myself. But there is a more raw, unabashed part of me that had been waiting to find this space for years, so it is a euphoric feeling to let it fully shine. It is probably not an accident that some of my favorite boxing trainers are also queer.

I largely agree with those who criticize the kind of health culture that is sold to women, trans, and nonbinary people. Yet I find that those critiques often don't leave space for queering that culture. Queer people are the most aware of the power and mutability of the body. Working out does not have to be in service of the kind of resilience that makes you a better worker or more obedient citizen. Generations of tomboys, butches, and transmasculine people who did not have access to hormones and surgical interventions worked this out long ago. Even when societal and medical gatekeepers could not see them for who they truly were, they showed it by molding their bodies to their own desires. Queerness, as I have come to understand it, is always future-oriented, always a state of becoming. Sometimes, I just have to "glove up" to make my own becoming more visceral.

I recently joked to one of my boxing trainers that in every class, I somehow end up standing next to someone incredibly strong who goes incredibly hard the second the lights come down and the music turns up. The trainer replied, "You know that you are that person for many other people, right? People say that you are so strong and badass." After class, their words stuck with me as I put my work clothes back on, buttoning an office-worthy shirt over my sports bra. This is why I come here, I thought. I come here to feel the edges of the strong person that lives inside of me, to grow into the version of myself that does not need to prove anything to anyone, to meld with the person that can find something of their own to hold onto, even in fraught spaces.

That person does not need an equation to tell them about their power.

GRAVITY

The words "gravity" and "gravitational" show up seventy-two times in my doctoral thesis.[14]

In 2017, a few years before completing that thesis, I had started a newsletter where I wrote long essays in first person, partly as a writing exercise, partly as a way to make sense of myself. I felt that writing about what I was thinking and experiencing on a regular schedule and for others' eyes would make me more honest and self-aware, and that it was not inappropriate for an aspiring academic to take up writing as a hobby. In its first issue, I declared that this newsletter would be "an exercise in honesty, an excuse for self-examination, and an incentive to keep in touch."[15] I wrote about gravity in the newsletter too. There, it was not, "We proceed to connect our work to realistic experimental systems and determine system parameters necessary for realizing condensate shells robust to the destructive effects of gravity," but instead, "I don't think I ever truly understood what gravity does, even when I wrote about it pulling on atoms in my publications and exams, until a yoga instructor tried to get me to flip upside down."[16]

Gravity is one of the four fundamental forces. It affects any object that has mass, and it causes any two massive objects to be attracted to each other. For instance, Earth's gravitational pull is keeping our feet on the ground, but we are also very slightly pulling the planet toward us. The sun and Earth have a similar relationship: their gravitational interaction locks our planet into orbit around the star. This is gravity as Isaac Newton understood it in the 1680s, and his ideas are still very effective in describing how we experience the force in our daily lives.

More than two hundred years after Newton, physicists began to understand gravity as something even more fundamental than a force. In 1915, Einstein's theory of general relativity advanced our understanding of gravity for extremely massive and dense objects—like black holes and neutron stars—and it explained the finest details of how planets like Mercury orbit the sun. It also introduced a large conceptual shift. As the physicist Robert Wald wrote in his seminal 1984 textbook on general relativity, "After one has learned the theory, one cannot help feeling that one has gained some deep insights into how nature works."[17] Wald is not being hyperbolic—general relativity deals with the very fabric of physical reality.

In Newtonian physics, space and time are assumed to be unchanging and mostly separate variables. Here, spacetime is flat and serves as a stage for objects influenced by gravity. Gravity acts between objects, but leaves spacetime itself alone. In the framework of general relativity, however, massive objects make spacetime curve. That curvature in turn makes the objects move along trajectories that we interpret as being shaped by gravity.

What is striking about this theory is that you could use it to correctly predict the motion of objects in our universe even if you knew nothing about gravity as a force—you could just watch all the valleys, hills, ridges, tunnels, and holes in spacetime and take note of how we might end up sliding, spinning, and falling near them. But Einstein's spacetime also responds to changes in the amount of matter and energy contained within it. For example, if you add more mass, spacetime curves more, like a taut sheet sinking underneath a heavy object. This spacetime is not a static stage on which physics plays out—it reacts as if it were another cast member.

Throughout the years, Einstein's theory of gravity has been incredibly successful. For almost as long as it has been part of modern physics, researchers have been concocting experiments to prove it wrong, and many such efforts are still ongoing. But whenever I've encountered cutting-edge research on gravity, the headline always ended up being about how Einstein was right after all. In 2024, an editor made me laugh by giving a brutally direct headline to one of my articles about general relativity: "Einstein's Theories Tested on the Largest Scale Ever—He Was Right."[18]

My doctoral thesis had nothing to do with Einstein, but space and gravity were still all over it. My first research project in graduate school—which would come to define much of my research career—involved taking a cloud of atoms, pushing them into a state close to absolute zero in which they were extremely susceptible to quantum effects, coaxing them into the shape of a spherical shell, then exposing the whole thing to the very weak gravity aboard the International Space Station (ISS). I was a theorist, so I never got to touch anything that later ended up on the ISS, but I solved an awful lot of equations about that cloud of atoms.

I discovered that unless you tuned every single parameter of the experiment just right, and unless you really did have access to outer space, gravity would destroy your ultracold atom shell every time. It would pull down the atoms on the top of the shell, deforming it into the world's coldest salad bowl. I wanted the atoms to tightly hold onto their unusual shell shape, but the relaxation offered to them by the curvature of spacetime almost always won out.

At the time, I was a regular at yoga, learning how to shape my body into Upward-Facing Bow, or Urdhva Dhanurasana in Sanskrit. It is a dramatic backbend. You start out lying on your back, knees bent and feet flat on the mat. Then, you place your hands by your ears and push your torso up, head off the ground, your belly button the top of a smooth arch.

The first time I successfully lifted myself up this way, I was overcome with adrenaline. I had been reluctant to try poses that I could fall out of, because I did not trust my body to hold me up or crash without disastrous effects. After a seasoned teacher gently coaxed me into this backbend, I had to reconsider this mindset.

Early in my practice, I'd wanted to control my body so well that it could temporarily defy and even defeat gravity. This did not get me very far. In fact, it set me back, as I'd leave class with bruised knees and, once, a contused rib. Even in common standing poses like Warrior I and II, the weight of my body caused pain in my legs, and my burning thighs had me standing all crooked and misaligned. I felt for the atoms in my equations because the atoms of my body also wanted to collapse. Teachers would gently correct me, only for me to tense up and fall out of alignment again.

At the end of most classes, I would lie in Savasana and feel my eyes well up with tears. The ubiquitous resting pose rarely brought me into a meditative state, but it was a reprieve from the feeling of fighting against an unbeatable fundamental force. A good enough reason to cry. And in those moments when the din of the world receded and I was forced to spend some time with only myself, I had other reasons for tears too. For instance, how stressful school was, how much I missed my partner as he studied in a different state, and how isolated I felt in the small Midwestern town of Urbana, Illinois. Sinking into the mat, I would feel the emotional knots in my body grow bigger and weightier. A force akin to gravity was pulling me down, tempting me to settle at the bottom of my depression.

When I was studying atoms, sometimes I would manage to plug in the precise experimental parameters and my equations would reveal the atoms resisting gravity after all. I never really worked out how to calibrate myself in the same way. I quietly kept struggling. I hid the heaviness within me by becoming someone who seemed to have endless energy, slept little, never said no to a project, and never ever skipped the gym. My days were composed entirely of routine, like existing on a factory line, and I created an illusion of control over my life by denying myself more and more things in the name of self-betterment.

Unless I was on vacation, I ate very little. I measured out my snacks with plastic measuring cups, and I prepped and portioned my meals with the kind of precision and rigor that I had previously reserved only for my academic work. I lost weight until my period stopped. I started eliminating food groups under the pretense of investing in my wellness, but really, I was just unwell. Everyone commented on my discipline and my thinness. This made me feel like I was on the right track. It made me feel like I was close to erasing all my problems, even though I was really erasing myself. I felt powerless, but I could still exert power over my hardened and knotted body.

Yoga had become part of my routine, and the wealthy white wellness influencers that I looked up to on social media presented it as a form of self-care that I absolutely had to do if I wanted to be healthy. I kept going to classes several times a week. I kept fighting gravity, and losing. A lesson loomed that I didn't want to learn: fighting the physics of my body, and maybe my body itself, would never work. As I became a friendly regular at the studio, more and more of my teachers implored me to listen to my body and work with the physics of it. Instead of hardening my body at the edge of each challenging pose, I had to rely on the curvature of spacetime to cradle me as I breathed in and out with intention.

Technically, the astronauts on the International Space Station are constantly falling toward Earth, except that deformation of spacetime around our planet turns their fall into an orbit. On my mat, I could sometimes admit to myself that I was also in a state of constant falling, then relax into that fall instead of tangling myself in anger and shame. As I got better at flowing without resistance in my yoga classes, my body started to reveal some of the sheer physical power that I would later fully step into as a boxer. But the seeds of that power could only be planted within me by learning that gravity does not have to be some sort of "final boss" supervillain that is on a mission to thwart my grand quest for perfection. If I could stumble and fall in yoga and forgive myself for it, maybe I could manage something similar in other parts of my life too. If I could trust myself to relax into challenging poses, maybe I could move through the rest of my days with more softness, more pliability as well. Practicing yoga did not make my life perfect, nor has it led to an unambiguously positive relationship with my body, but throughout some of my hardest years in graduate school I learned some of my most transformative lessons on the yoga mat.

Years later, I found myself in a very different kind of yoga class, surrounded by soft bolsters and heavy blankets, a little high and filled with love for my friends and my partner practicing alongside me. In a very serious tone but one not lacking in kindness, the instructor explained that we would hold all poses for three to five minutes, draping our bodies over many supports, and letting gravity tug not just on our muscles, but the connective tissues in between them. This is yin yoga—a slow-paced style that incorporates concepts from traditional Chinese medicine. She encouraged us to find our edge and stay there, never fully crossing into pain, but not fearing to approach it either.

I thought about those atoms again, about deriving an expression that perfectly balanced their quantity, the strength of the interactions between them, and the size of the shell they were trying to make to keep themselves at the edge of collapse. Three to five minutes at a time, I inhaled and exhaled through my nose, trying yet again to trust gravity to stretch me out, straighten me, and unknot me. This time, I was not at all afraid of it taking away my power.

The next day in boxing class, when I threw my first hook, a new kind of soreness reminded me of how I had communed with gravity the night before. Gearing up for another punch, I decided that I was grateful for the soreness, then let my fist fly. As a student of mine once explained in a well-executed assignment, a combination of Newton's First and Third Laws of Motion caused my gloved hand to encounter resistance when it hit the bag, and a small reverberation traveled up my arm. The knots in my shoulders, softened by the previous night when I had let myself give up the fight, rattled loose.

"Give me your best power shots," the trainer yelled from the podium. And within my strong, queer body, I had so many.

I started teaching ninth grade physics during the COVID-19 pandemic in 2020, so many of the attempts that I made to connect with my students as a first-year teacher were mediated by a screen. They were new to high school, and I was new to teaching. Together, we were learning how to be in our new school and claim it as our own despite not being in the building every day. Some of my students attended this school remotely for a full year, only beginning to attend in person as sophomores, after the virus had been declared less dangerous and the school stopped offering remote classes. During that first year, I experimented a lot with lesson and assignment formats, of which Sports Week was my most successful experiment. Its objective was for students to apply the concepts of momentum, impulse, angular momentum, Newton's laws, and two conservation laws to situations that arise in sports. We ended up doing much more.

During our video call classes, the students told each other, and me, about their favorite teams and athletes. They bonded over their presentations, even when that meant overcoming the fear of speaking not just publicly but publicly over Zoom. After weeks of solving equations while bathed in the harsh glow of laptops and iPads, everyone enjoyed the reminder that the real world, where physics happens all the time, can be fun. When one student thanked me for coming up with Sports Week, I cried at my kitchen table turned teacher's desk.

The next year, with school now precariously in person every day, I decided to go big and turn the end of Sports Week into a student-led "sports film festival." I brought popcorn, and students screened the videos they had made about transfer of momentum in their friends' soccer games, or the impulse a human hand can exert on a boxing bag. There was lots of "oohing" and "ahhing," and a lot of greasy hands shooting into the air to pose physics questions to their peers.

Sports Week worked so well, at least in part, because it was physics you could feel rather than just physics you can learn.

This is my Sports Week project.

PART 3

INTERACTIONS

CHAPTER 8

ULTRACOLD

In January of 2020, I thought I had written a great sentence: "The coldest places in the universe can often be found in a room on your local college campus, and they are likely controlled by a graduate student."[1] The magazine editor that I was working with was mostly unimpressed.

I was writing an opinion piece about my favorite branch of physics, "atomic, optical, and molecular physics," a gig I'd landed by attending a science communication workshop, and the editor was stuck having to shape it up. I had not yet finished my PhD, my writing experience was very limited, and many of my peers were the very graduate students referred to in that sentence. Every day, they were tending to extremely cold atoms, and I was their theorist friend who wanted to discuss it. That sentence, the nod to my friends, and my fledgling writing voice eventually all made it to print. The piece ran in *Scientific American*—my first magazine publication. It was an early, tentative taste of my future.

Really, I would not have a writing career if I had not first become obsessed with ultracold atoms. When I say ultracold, I mean only a few billionths or trillionths of a degree above absolute zero. That opinion piece opened the door for me to start reporting on physics and eventually transition into doing that full-time. Reporting on this type of research is still a big part of what I do: if you type "New Scientist ultracold" into Google, dozens of articles will fill your screen, mostly with my byline attached. In 1997, decades before I was hired there, *New Scientist* ran a headline that read "Just Cool It—Atom-Chillers Win the Ultimate Accolade." The accompanying article announced that Claude Cohen-Tannoudji, Steven Chu, and William Phillips had won the Nobel Prize in Physics for working out how to use lasers and magnets to trap and radically cool atoms.[2] I was about to turn six, a few months away from starting elementary school, and the closest I got to physics was the occasional conversation with my father about his work as an electrician. Every memory I have of my life before school—which, incidentally, was also a time when physicists were chasing ever more extreme levels of coldness—is a distinctly warm one.

I remember eating my grandfather's homegrown watermelons on rocky beaches, picking sun-soaked cherries with my dad and my grandma and then baking them into a piping hot strudel, going to kindergarten, located next door to my parents' apartment, and running home afterward through a sunny playground. My brother was born during the searing hot June of my fifth year, so my first memories of him are also warm, such as my mom fanning his large baby head in the back of the car on the way home from the hospital.

There are winters in Croatia, even on islands like the one where I grew up, but those early childhood years replay in my mind as a succession of endless summers: the air heavy with humidity, the sun bright against the muted blue tones of the Adriatic, the dirt from my grandfather's vineyard sticking to my sweaty hands. In one picture from those years, I am standing on the road leading to the vineyard, wearing a straw hat. My face is framed by loose chocolate-colored curls, and I am looking into the camera sweetly, without an ounce of reserve.

In the seventeenth century, physicists thought that warmth and cold were two separate substances. Also popular was the idea of *primum frigidum*—a substance that was as cold as anything could possibly be, a kind of anti-fire that made other objects cold by touching them. Robert Boyle, an influential chemist of the era, eventually disproved this idea based on his experiments, and ultimately the realization caught on that cold is not a thing but rather the absence of a thing—the absence of heat.[3]

To make something cold, then, you must take away its warmth. To make it ultracold, you must excel at that subtraction. But heat naturally "wants" to spread—as demonstrated when summertime heat makes contact with a glass of ice water, or when the heat of steaming wintertime tea spreads itself across the chilly atoms of the kitchen air. The branch of physics that worries about why heat wants to do this, and how this can be reversed or prevented, is called thermodynamics. And thermodynamicists are still out there, studying it.

In 2001, when I was ten, another Nobel Prize was awarded to another three physicists for another feat of ultracold physics.[4] I understood a little more about atoms now—still mostly thanks to my father, but school was also helping. And I knew more about what happens when warmth starts slipping away from a person. I was learning about that because I was seen as a girl, and society deemed it crucial that I maintain the sense of friendliness, softness, and warmth that I'd had in all those sunny, watermelon-sticky, cherry-sweet memories.

In Croatian, you can call a person cold, but native speakers tend to be more specific. "Stuck up" and "conceited" are favorites, and I became the target of those phrases before I even hit puberty. There are more imaginative ones too, such as accusing someone of "acting as if they are important," or calling them "inflated," as if they'd swallowed air to try to become bigger and take up more undeserved space. There is also the Croatian word for "stiff," which invokes a starched collar. In kindergarten, boys picked on me for allegedly making up big words to sound important, but increasingly, this wealth of barbed language started to come from adults in my life as well.

As I approached my teens, aunts at family gatherings and birthday parties gave me sideways glances for talking too much and too confidently. Uncles and grandfathers either dismissed me or—as my maternal grandfather would do well into my college years—spoke down to me until my cool cracked and I cried. In school, I struggled to make friends, prompting my father to share that my mother had been similarly unpopular when she was young because she too had been seen as stuck-up—what my American peers would later describe as cold.

He wasn't yet ready to use any word that would translate as "bitch," but I understood what he was trying to say—and that he wasn't taking her side. When my mother showed me photos of herself as a teen, she would critique her posture and clothes, rarely taking her own side herself. In college, I would catch myself sharing a childhood photo on Facebook and wondering in the caption what had happened to that sweet, cutesy kid. That kid, I knew, had grown into what our culture viewed as a hard-to-handle, self-important, loud young woman who lacked a motherly instinct—a feminine softness and a warmth that should have been innate. That kid became a teen who instead of conforming themselves to a standard started to embrace a kind of iciness, soon fully committing to a cold and rational manner of speaking, and an unbecoming trust in their intellect.

In high school, and then again in college, I learned some basic thermodynamics and found it unexciting. My classes made it feel like a solved discipline, a dusty nineteenth-century foundation for more exciting courses of study. But in college I had a cold, steely resolve to attend every seminar, colloquium, and invited talk that was offered, so it was inevitable that I would eventually hear the words "ultracold" and "quantum" in the same sentence. Eventually, someone put an equal sign between them—and I found that revelatory.

In my memory, it was a famous French physicist speaking to a group of us undergraduates over a free dinner who first explained to me that you can use lasers to make atoms ultracold and reveal their quantum character in the process, but I have never been able to confirm his name. The event is seared in my memory, but not the records of my college's physics department. I am more certain that, just a few months after that first revelation, I shook hands with William Phillips, one of the 1997 Nobel Prize laureates. He had been at the National Institute of Standards and Technology (NIST) for years at that point, and he gave me an official NIST card, printed with the entire International System of Units of Measurement, for me to carry in my wallet. I never worked in a lab, so I never really needed it, but I kept the card and often thought of his work as ultracold atoms overtook my scientific imagination.

In his Nobel lecture, Phillips said that, at the beginning of his work with laser cooling, he did not fully appreciate all the "exciting possibilities of what one might do with laser cooled atoms."[5] By the time I got into the field more than fifteen years after that lecture, those possibilities were clearer, but not any less exciting. For one, the colder you go, the more quantum effects that you would never see at room temperature become prominent, eventually taking over. They are not just odd and counterintuitive, as quantum effects tend to be—the ultracold quantum realm also reveals a lot that is truly new and surprising.

Some of the exploration that is happening there now has ramifications for the field of materials science: it could help us design materials that conduct electricity more efficiently, or lead us to new kinds of computers that can solve currently unsolvable problems. But some of it is also exploration for exploration's sake and has already uncovered new phases of matter and physical phenomena. Wolfgang Ketterle, one of the three 2001 Nobel Prize in Physics laureates, told me in an interview several years ago, "As a general principle, if you can reach lower temperatures you are bound to make discoveries. If you live in a desert, you could not discover ice unless you build a refrigerator. That's what we did."[6]

I was interviewing him for a feature that had the working title "Conquest of the Cold," imagining the ultracold realm as the next frontier for contemporary physics. I am glad that my editors dropped that framing before the piece made it to print—extremely low temperatures are not so much a frontier for colonization but rather a place where our understanding of quantum meets its edge, forcing us to think more imaginatively. Solids that are also fluids that flow with zero viscosity, magnets with six types of "poles" instead of two, pools of atoms that can switch between being different types of particles, odd sound waves that are tiny equivalents of undulations of spacetime—they all reside in the ultracold realm.

Several physicists have told me that we are so new to the ultracold game that we sometimes do not even know how to properly observe and identify exactly what an ultracold system is. In conditions that are more intuitive to creatures like us, who naturally run at about 300 degrees Kelvin above absolute zero, we have a useful shorthand for what temperature is. For a collection of atoms that jiggle in place with some known average velocity, more motion means a higher temperature, while less motion correlates to a lower temperature.

But under ultracold conditions, that shorthand starts to break down, and other properties—like entropy—become relevant. In 2022, when I reported on an experiment with an ultracold magnet made from ytterbium atoms, my interview with the project's theorist helped me understand just how difficult it can be to determine just how cold an extremely cold object is.[7] If you discover a radically new form of matter in the ultracold realm, he told me, you likely don't know what properties its atoms have at every given temperature, so it becomes really difficult to build a device that could identify its coldness in the manner of a thermometer.[8] The strange magnet at hand was probably colder than the estimated temperature that his team had published, he conceded, but because they couldn't measure it directly, or didn't really have a good idea of how to even go about that, they'd had to play it safe by comparing the magnet to theoretical models that could be simulated on a computer. The computer could only simulate a crude version of the thing, but it was, at the time, their best available thermometer.

My father always knew how to identify women who were insufficiently warm, who had strayed too deeply into the cold, rational, aggressive realms traditionally reserved for men. His sociopolitical thermometer was never off. If a women's rights protest showed up on the news when I was young, he would always tell me that women would never accomplish anything by being so offensively angry. I often thought of this during my last two years of college as I was becoming a more outspoken feminist—or at least what counted as an outspoken feminist on Facebook and Tumblr in 2013—while dating a man who wanted me to be less heated about it.

He was another physics student, another aspiring physics professor, another ultracold atoms enthusiast. At one point, we both approached the same faculty member about working in his lab. This professor told me that I was "not a typical physics student" and that I should take an electronics course to learn how to operate "a machine other than a microwave." My then boyfriend, on the other hand, used his limited past experience working in a biophysics lab to secure a paid position.

He built devices that were then used to create some of those coldest spaces in the universe, and he loved to talk about it. When we lived together, he would bring home gaskets and other components from the lab and quiz me on their function. I rarely knew, so he would lecture me on it as I made us dinner. Eventually, I joined a research group too, one headed by a female professor with a reputation for being difficult.

She let me mess with a simulation of an elongated cloud of ultracold atoms that was being kicked in a very specific way and, to my delight, she also taught me a little bit of the kind of quantum theory that was beyond the scope of the classes I was then taking. She told me stories about having been treated poorly as a professor when she was pregnant, and about the unprofessional ways her colleagues responded to her ideas at seminars and conferences where she was typically one of the very few women in attendance. She was not really a difficult woman—just an outspoken one. My conception of my gender has changed since college and I do not think of myself as a woman anymore, but I still think of that professor as my elder: a woman who spoke up to men instead of offering them nurturing and patience, and who also had an interest in the ultracold.

In a 2017 survey of 4,573 American adults, the Pew Research Center found that the majority thought the traits society valued most in women were physical attractiveness, a nurturing personality, and empathy. For men, the top answers were more diverse and included honesty and morality, professional or financial success, ambition or leadership, strength or toughness, and a good work ethic. Survey participants largely agreed that men and women are fundamentally different, but there was no consensus on whether that was because of biological nature or societal influence.[9]

Researchers still disagree on how to disentangle the two, citing both the evolutionary benefits of nurturing qualities among child-bearing people and the nearly impossible task of removing cultural influences from any study. Some have turned to brain scans to show that those assigned female at birth seem to feel more empathy when confronted with others' pain.[10] Others have studied who is more likely to donate to charitable causes.[11] And countless think pieces have been written about the impact of telling women to smile more, or the hidden burdens and value of their emotional and domestic labor.

In my college years, I was incensed by my gendered experience of the world. This, in turn, incensed my then boyfriend. Unlike my father—who wanted me to be softer, warmer, and less angry—my boyfriend wanted me to be more resilient and self-sufficient. He would compare me with his mother, a woman with whom he was often in conflict but who had, in his eyes, earned the right to be a feminist by arguing rationally and not asking for anything he deemed unreasonable.

The first time I met her, she told me that it was no wonder that young women were being sexually assaulted on college campuses when they walked alone at night and wore short shorts. My hopes of finding a kindred spirit in her alleged feminism vanished. Yet, I wanted to win her over, so I made us all bread. I vividly remember her snickering when I got flour on my shirt while I was kneading the dough. I was embarrassed and so was the boyfriend. I had not been cold enough to argue with his mother about sexual assault, and my attempt at being nurturing had been too sloppy. I could tell that my ex was thinking what I often thought during our life together—that this was just another example of me not being able to set my thermometer right.

He wanted me to be tough and resilient, and to stop being so despondent and whiny about the plights of my then gender. But he also wanted my labor: in the kitchen, around the house, and as a listener and cheerleader when he'd had a bad day. The trouble was, I never did the dishes correctly, never folded the towels in our bathroom right, and rarely said the exact right thing when he was in distress. I felt like I was in a special program for becoming a "normal girlfriend," forever stuck at one of its early steps, the next one constantly shifting, its temperature always impossible to predict.

He did not need me to be warm all the time, but I had to be able to turn it on, and to turn it on just right, when he deemed it necessary. And my coldness? That had to be controllable too, deployed only when a practical, rational approach was needed to solve something, like how to do homework. An approach as utilitarian as the cooling apparatus he was building in the lab.

Here's how you cool a warm, jiggling gas of atoms. First, you put them in a chamber made of metal and glass that has no air inside it. This gives you more control over the atoms. Then you hit them with lasers. The photons that comprise the laser beams barrage each atom. When hit, an atom absorbs a photon, then re-emits it, and the re-emitted photon takes away some of the atom's momentum. Momentum is directly proportional to velocity, so decreasing the atom's momentum slows it down and makes it colder. At this point, you will also want to add specially designed magnets to your setup and use their forces to keep the atoms from speeding up and escaping the nascent ultracold region.

You can then decrease the atoms' temperature even more by using microwaves to do something akin to blowing on a hot cup of coffee. Blowing adds energy to the region and helps the warmest and fastest molecules escape, leaving behind only those that are cooler and more stationary. Each cooled atom is a quantum entity, so it cannot have just any amount of energy—there is a special list, or ladder, of energy values that it can settle into. Researchers use bursts of microwaves to push atoms up the ladder and eventually out of the ultracold atom cloud, lowering the overall temperature of the remaining atoms. What's next is up to you—you can control your ultracold atoms with extreme precision using the tools that you already have: lasers, magnets, and microwaves.

The first time you hear it, the idea that the only way to study some phenomenon or create some type of matter is to use ultracold atoms can seem bizarre. It becomes less odd once you understand that a warm atom can never be controlled as tightly. In recent years, this principle has even carried over to chemistry, as researchers have learned to incite reactions between ultracold atoms or molecules, then halt them and see intermediate products that would be invisible at room temperature.[12]

Outside of physics, everyday language also implies a connection between coldness and control, like when we call people hotheaded for acting impulsively. I have always gravitated toward discipline and self-control, which has certainly contributed to my perceived coldness. But it has also been an effective defense mechanism. It made it more difficult for others to manipulate me by wedging themselves into space left by an emotional outburst.

In college, personal conflict would draw tears and rage out of me, but by the time I completed graduate school, I was a lot more adept at shrugging things off. With a flat inflection, I would claim that most things are simply not worth getting upset about. I still largely operate this way, sometimes to a fault. As a child, my confidence and lack of deference were misinterpreted as a grating kind of coldness, something lacking from my character. Ironically, I grew into a kind of coolheadedness that is a lot more literal, that I cultivated on purpose.

At twenty-three and at the edge of graduating from college, undoubtedly an adult already, I was caught in a disintegrating relationship with that other ultracold atom enthusiast. Our hopes of a shared future as two hotshot young physicists had dissolved into something sticky and viscous that was not really holding us together but was certainly keeping us from wading too far apart too quickly. I felt like I had lost an unforgivable amount of control over myself, and I felt like I had failed. I'd failed to find the sweet spot between cold rationality and warm domesticity, between providing my ex fun intellectual challenges and yielding when our heterosexual power dynamic felt endangered.

In our relationship, it was a problem when I cried, but it was an even greater problem when I expressed a masculine-coded independence—that lack of feminine warmth that my father had warned me about. At one point, my then partner started sleeping on the couch to, as he said, avoid dealing with the mere possibility of my sexual advances, potentially the most embarrassing punishment for me overstepping the crude bounds of the traditionally feminine. Whether making cupcakes or hosting dinner parties or changing my hair or coaxing him through painfully mathy homework, whatever I tried to do to warm myself up and dull my edges was just not enough to prevent me from ultimately becoming the cold, stuck-up, starched-up, self-inflated bitch that I feared hid in my genes.

In social psychology, all the perceptions one can have about another person, or group of people, can be placed on a coordinate system with two axes: warmth and competence. "Warmth" is equated with trustworthiness and friendliness, while "competence" stands for capability and assertiveness; warmth says something about intent, and competence measures the likelihood of that intent leading to action.

This theory, called the "stereotype content model," was put forward by a team of psychologists in 2002.[13] They based it on empirical data, and it has since been corroborated by many, many studies. In 2018, Susan Fiske, one of the model's originators, wrote that "these Big Two dimensions converge across survey, cultural, laboratory, and biobehavioral approaches. Generality across place, levels, and time further support the framework. Similar dimensions have emerged repeatedly over the history of psychology and in current theories."[14]

Different quadrants of this coordinate system correspond to different stereotypes. The ideal citizen lives in the quadrant of high competence and high warmth. The quadrant of low competence but high warmth is populated by people we see as deserving of pity and a very paternalistic kind of care, like the elderly. In contrast, researchers find that the mix of high competence and low warmth describes technical experts and the rich. People in surveys report feeling envious and resentful of this group. I suspect this quadrant had been my home ever since kindergarten and that "big word" incident.[15]

I learned about this model in 2015, when I started dating a social psychologist. We were both in the early stages of PhD programs, and we shared a love of heavy metal and greasy 1950s hairstyles. He was researching how people's perceptions of politicians as warm and competent influence voting patterns, and I was doing my research on the hollow spherical shell made from ultracold rubidium atoms that would eventually be created aboard the International Space Station. We discussed our respective lines of work from time to time, but mostly we watched *The X-Files*, listened to music on vinyl, cooked together, slept in late, drank cheap whiskey, and ate homemade bread on his stoop. We made loose plans, played a lot of mornings and afternoons "by ear," and at night we kept extending each other invitations to come over for one last cup of coffee, even when neither of us actually had any in their kitchen. We lived in Urbana, Illinois, where it could get so cold that classes had to be canceled, but this is another era where I only remember warmth.

Graduate school would ultimately prove to be immensely difficult, but during that first year my classes were an exhilarating challenge, my ultracold atoms research was a never-ending source of possibilities, and this new relationship overflowed with moments of care and gentleness. Under an ever-cooler exterior, I felt something like the beginnings of a thaw, like there was a type of warmth at my core that I had never accessed before.

On rare bad days, I worried that the psychologist would try to place me on that "warm and competent" chart and find my coordinates disqualifying. But he never did, and I realized that I could adjust them without having to trade being resented for being pitied.

I also discovered that my analytical, practical skills could in themselves foster a kind of warmth. First, I got involved with launching a peer mentoring program for physics students and eventually became a member of its leadership, at least in part because I had a knack for planning retreats and starting events on time. I kept spreadsheets, sent emails, and made PowerPoint slides with actionable, well-researched tips for navigating the world of academic physics.

This led me to consulting for a national network of similar programs and co-organizing several of its conferences, which remain some of the most intentionally transformative and inclusive spaces I have ever encountered. Later, I was on the board of a graduate employees' labor union, mainly dealing with its finances, and I also got an advocacy group for women and gender minorities off the ground. By the time I graduated, while some of my younger peers knew me as a "cold atoms" theorist, most knew me as someone who could be a resource—or even just a shoulder to cry on—when they felt confused, mistreated, or low.

I hadn't realized how much I'd changed until a roommate told me that she remembered me from an event that we'd attended while still in college. In her memory, I had been the coldest and most intimidating person in the room back then. It hurt my feelings to hear that, but there was also truth to it. At the time, I'd been guarded, untrusting, and held tightly onto my defense mechanisms. Similar to how I thought of myself when I was a preteen, I understood that someone was likely to object to my lack of sweetness and deference, yet I doubled down on being icy.

My personal hurt aside, it is objectively true that in most workplaces, including academia, women—and people perceived as such—are expected to lean into their supposedly innate desire for caretaking and do extra work that is not expected from their male peers. A 2024 report by the consulting firm McKinsey & Company and the women's campaign group LeanIn.Org found that women were not just overwhelmingly taking on menial office tasks, like stocking snacks or taking notes, but were also supporting the well-being of their colleagues and spearheading efforts for diversity, equity, and inclusion. In one national study of over three thousand engineers, both white women and women of color reported doing about 29 percent more office housework than white men.[16]

The 2024 report found that this significantly contributed to feelings of burnout, prompting many women to consider leaving or downgrading their roles. Less than a quarter of the 423 organizations that participated in the study recognize this type of work in formal employee evaluations, thus rendering it virtually invisible as it relates to job promotion. Academia is no exception—at top schools in the country, female faculty are most likely to do "service" work without any of it significantly counting toward their cases for tenure and other promotions. The same is true of other minoritized academics. For instance, one 2012 study found that academics of color spent three more hours per week on service tasks than their white peers.[17] Though I never made it far enough in academia to experience this firsthand, at my graduate school alma mater, I witnessed women faculty—especially minority women—be inundated with service projects and diversity-related grant proposals.

Chanda Prescod-Weinstein, the first Black woman to hold a faculty position in theoretical cosmology or particle theory, writes of her own experience, "The people who get most swept into this academic housework are people who aren't white cis men. Also unsurprisingly, the people most likely to win awards and generous praise for doing this work are white (men). We people of color and white trans/genderqueer/cis women and nonbinary people do the academic housework that helps keep academia running. We are not always paid for this work, and when we are, it's often not enough. And we are punished for it when it comes time to apply for jobs and tenure."[18]

As a graduate student, most of the people I organized with were women and queer people. We were like a junior varsity team for the queer and female faculty who mentored us, doing the same kind of invisible work but with fewer titles. When we tried too hard—such as asking that some spaces and events be closed to cisgendered men or going on strike for better working conditions—the university as a whole, and some male faculty members as individuals, pushed back with no lack of paternalism. Ironically, that college boyfriend prepared me for this wobbly walk on the thin line between being a confident voice for change and being not so cold as to be immediately off-putting. But it made a world of difference that I was no longer navigating this alone.

The personal warming-up project that I undertook in graduate school was largely done on my own terms, and in community with like-minded people. Only by sitting with my peers who wanted the same thing that I did—to make the graduate school experience better for all of us—could I recognize my own flaws and warm myself up in a way that did not feel self-destructive. They helped me become a better listener, less of a black-and-white thinker, and someone able and willing to hold space for discomfort and the uglier parts of personal growth. We were all tired, and we were all frustrated, but we could look at each other and be affirmed in our shared struggle. In my own way, at my own pace, without negating myself, I felt like I was moving closer to that aspirational quadrant of high competence and high warmth.

Relying on the free labor of graduate students to bring about better working conditions and more inclusion is yet another institutional failure of academia, but I do not regret having been part of it. It was affirming and it also made a difference. I'm certain of this because current graduate students still sometimes call me to ask for advice, on the recommendation of their older peers who remember me. When I go to physics conferences as a journalist in search of physics news, I often hear about retreats, union actions, and new student groups at my alma mater. In 2025, during one such conference, at an alumni reception, a current graduate student told me they had "heard the legend of Karmela," and I nearly spilled my glass of water on the friend I had been speaking to. The next day, I reunited with the group of former graduate students who had ushered me into that peer mentoring program. It was its tenth-year anniversary. Sharing deep-fried mushrooms, noodles, and bubble tea, we laughed so much and had such warm conversations that the ten intervening years, and the geographic distance now between each of us, started to feel insignificant. When the second administration of Donald Trump started to systematically dismantle all efforts to make American science more inclusive, these were the first people I wrote to. But even in times of crisis, it is encouraging to know that our efforts toward building a better future have taken root, even if in relatively small ways. As the activist Grace Lee Boggs wrote, "Dramatic and systemic change always begins with 'critical connections,'" and I can see those growing within this microcosm that was once my everything.[19]

I wish I could say that, after I left academia, no one ever again made me feel like I was too cold, or accused me of being pretentious, or conceited, or stuck-up. In fact, I have never had a job where some senior person did not find it necessary to pull me aside and chastise me for not being friendlier, sweeter, warmer. I also wish I could say that these repeated incidents did not upset me or remind me of every other time I had been similarly reprimanded at ages ten, twenty, or thirty. But I let go of the hurt more quickly now. And I am better at diagnosing what people really mean when they call me disagreeable and cold.

Sometimes what they are objecting to is my drive and expertise. Sometimes it is that they have misread my gender and expect the kind of acquiescence from me that is traditionally, cruelly, expected of women. Sometimes they have read the masculine in me correctly but want to punish me for it. Sometimes the issue is that of control, and how tightly I can hold on to it when it comes to my work. Sometimes I am actually at fault and really do have to learn and recalibrate.

It helps that I continue to not be alone. Instead, I am surrounded by friends whose care feels accepting of a range of temperatures. I have been a firm but loving older sibling figure to many of them, workshopping difficult text messages to family members, dropping off homemade meals during depressive episodes, going on long walks on lonely holidays, planning elaborate surprise parties, repeating "You did nothing wrong" and "I will always be here for you" over and over until the torrent of tears on the other side of the phone abates to a drizzle.

I may not be a person whose presence bathes everyone in warmth like that psychologist boyfriend (whom I ended up marrying), but I have learned that a calm, coolheaded patience can also help grow the kind of interpersonal bonds that help us stand a chance of building a better future. I know this kind of patience is not exactly what was radiating out of that curly-haired kid in the straw hat, but I believe something like it was always within me. It just took time to find an environment where I would not be pelted with expectations or forced to stay in my place, or where the air that could sustain the fire within me had not been removed.

About a month before I started writing this chapter, I was on a call with another team of physicists who told me they had done something unusual with a cloud of ultracold atoms: they had created quantum matter with negative absolute temperature, or temperature beyond absolute zero.[20] Quickly, they assuaged my shock: these atoms were not colder than absolute zero but rather, in a quirk of how temperature is defined in quantum mechanics, hotter than anything that has a positive temperature. If you put one of these atoms that have what seems to be an impossibly cold temperature next to a room-temperature atom, heat would flow from the former into the latter, a researcher patiently explained. Later, an editor, while reading the draft of my article about this experiment, bemoaned how complicated thermodynamics gets in the ultracold quantum realm—but they also said they were not at all shocked to see my name on top of it.

CHAPTER 9

DIFFRACTION AND SPECTROSCOPY

Diffraction: the way light bends as it moves past an edge of an object or through a narrow aperture.

Spectroscopy: the study of electromagnetic spectra, examining phenomena such as how much red light or how many microwaves have been emitted by a star, a flame, or some other object.

Taylor Swift's 2022 song "Vigilante Shit" opens with eyeliner: "Draw the cat eye, sharp enough to kill a man."[1] The song's narrator is something of a femme fatale, and once you have that black flick at the corner of her eye to hang your imagination on, the rest of her materializes more easily. Swift is known to have loved Tumblr in the 2010s, like many other millennial women, and the site was full of memes about sharp eyeliner as a weapon against the patriarchy. Big beauty brands were getting in on the feminist framing of makeup too, with CoverGirl launching a flashy #GirlsCan ad campaign in 2014 that featured several influential celebrity women. Later, in 2018, the company put out ads featuring "Girls Who Code," which showed teens in lip gloss and the occasional line of what looked to me like C++ code that I had also once tried learning. I learned a lot about gender and politics from Tumblr back then, including from commentary that endorsed or criticized those eyeliner memes. When "Vigilante Shit" was first released, Swift's lyric reminded me of how different I, and the

Internet, used to be. But some things persisted—I am still not a Swiftie, and Taylor and I are both still committed to cat eyeliner.

I started wearing eyeliner at the beginning of high school, not because I wanted to be a femme fatale, but because I was slightly confused. My identity at the time was built around being a heavy metal fan, and I was trying to negotiate the genre's fashion. In 2000s Croatia, the heavy metal look was still largely modeled after the 1980s stylings of Iron Maiden and Judas Priest. There were skinny pants, leather, spikes, and band tees, but those were mostly for slouchy, long-haired boys who liked beer and cigarettes. Femininity had never been intuitive for me, but when I didn't dress like one of the metal boys, the only other option seemed to be to pattern myself after some sort of hot goth girl. I was enough of a nerd and a snob to insist that metal and goth were not equivalent, but the goth look was more compatible with the female gender that people ascribed to me. It also did not have to look natural, so it hid my awkwardness around expressing girlhood. I got into fishnets, long black skirts, and—notably—eyeliner.

I no longer dress to match my taste in music, but femininity remains unintuitive to me. Though I don't dare call myself butch, my daily uniform of slacks, a button-down, and a bandanna tied around my neck betrays an affinity to a masculine mode of queer expression. But I also have my "standard face:" dark, faux-laminated brows, blush so abundant as to make me look sunburnt, highlights that make my cheekbones shine like chrome, lips overlined into a neat circle, and bright eyeshadow in greens, blues, and pinks. Sharp black cat eyeliner is a staple of this face too, so much so that if you run into me on a workday and find me without it, you can assume that something's gone terribly wrong.

In the summer of 2024, I wanted to report on a new imaging technique called "event-responsive scanning transmission electron microscopy."[2] I messaged my editor:

> Electron microscopes shoot electrons at samples, which can produce damage, so they are not suitable for fragile materials and biological

> tissues. This paper [that I propose to write about] is about a new way to use an electron microscope where instead of shooting electrons for some preset amount of time, the microscope reacts and adjusts to the sample in real time. It basically shuts off once it has seen enough and doesn't cause as much damage. The researchers tested this out on a thin sample of biological tissue and on a mineral sample, both of which are known to be very sensitive to the deleterious effects of electron beams.[3]

An experiment like this is exciting for many reasons: there is the wow factor of how small electrons are, the remarkable fact that we can control them, the promise of what electron microscopes may show us in the future. But I also get stuck on the notion of there being something inherently violent about how these devices work. Viewing an object, or its insides, through an electron microscope always requires a bit of an assault, as you must barrage it with particles. Moreover, electrons are not the only particles that lend themselves to this process. To see inside of objects you can also, for example, hit them with the more massive neutrons.

While the best conventional microscopes let us see objects hundreds of times smaller than a grain of pollen, electron microscopes can image objects with atomic resolution—they produce pictures in which you can tell atoms apart. This happens in one of two ways.

In the spectroscopic approach, you hit the object with an electron, which knocks into some of the other electrons in the object and causes them to change their energy levels. That shuffle promptly results in some of the energy being released as X-rays, which the microscope then detects and turns into an image. In the second approach, which involves the phenomenon of diffraction, you hit the object with electrons, which hit the object's atoms or pass through the spaces between them and emerge onto a detector. This detector measures an assortment of the properties of these electrons, then uses those measurements to construct an image of the atoms inside the object. In both approaches, the object faces a series of potentially damaging collisions.

In these and similar experiments, a physicist's job is to direct the collisions and analyze their aftermath, both of which are tricky tasks. It is especially important to choose the right kind of particle—be it an electron or a neutron—and to give it the right kind of energy; otherwise, it may not penetrate deeply enough into the object, or the object may fail to respond, resulting in useless images or no images at all.

The design of the first electron microscope is credited to the German physicist Ernst Ruska in 1933.[4] He would receive the Nobel Prize in Physics for this work more than fifty years later, but the Germany technology company Siemens started making electron microscopes only a few years after he first finalized the design. However, for a while it seemed like the device would never be practical, let alone the mainstay of materials science labs that it is today.

It seemed impractical because in those early designs, the electron beams often destroyed the objects meant to be imaged, burning them upon impact. The trade-off between damage rendered and information gained was too high. Researchers who inherited Ruska's work later found ways to mitigate that damage, and electron microscopes became widely used, but the issue persisted to a certain degree—or I wouldn't have been pitching a story about it in 2024.

"The 'makeup is empowering' narrative is bullshit if you think your cat eyeliner will truly ever be sharp enough to stab through the eyes of a catcaller, but I don't think CoverGirl's sell was a total smoke screen," wrote the beauty writer Arabelle Sicardi in 2014. Their piece reckoned with how often makeup is used to cover up the marks of domestic violence and CoverGirl's decision to not terminate a partnership with the National Football League after a video surfaced online of one of its players attacking his fiancée.[5] Sicardi continued, "I bought into the fantasy of beauty products when I was being abused because I found them to be comforting, momentarily palliative. . . . More than anything, the ritual of beauty—the familiar act of putting on the cream, drawing the lines—made me feel like I had control over something."[6]

I was in my twenties when I read this piece, and it rattled me. I am lucky to have never been physically harmed by a partner, but there was something else making me uncomfortable here. At this point, I had moved past the "hot goth girl" style, and the idea of makeup as feminist rebellion had steadily grown on me. I read online screeds by self-proclaimed feminists, and I wore heavy eyeliner and red lipstick to my college physics classes because I wanted to think of myself as untamable and independent. Sicardi's piece forced me to reexamine whether it was really that simple.

Why did heavy makeup appeal to me to begin with? Was it about fear and control for me, too? On days when I am particularly stressed about matching the color of my eyelids to my shirt, and my highlight to my belt buckle, these questions still feel extremely potent.

My makeup budget is bigger now. I use more products, and my morning routine is notably longer. Layers of priming, bronzing, and blushing give me a more polished face, and more of that sense that I'm exercising control. The way I draw my face does not make me more conventionally beautiful, nor does it read as natural. But it does signal that I have taken control over my features—I make them big, colorful, and visibly artificial because I am in charge. It also puts up a literal barrier between me and the people who look at me. Some of my coworkers have never seen me without my standard face. In situations of conflict, I feel calmer knowing that I have already unilaterally decided some of what they will see.

Most of what I know about makeup I learned online, and the Internet has always been vast enough to cater to my idiosyncrasies. I never learned how to do "clean girl makeup," because I was watching video tutorials of drag queens, former club kids, and influencers who were trying to represent specific subcultures. I grew up with '80s heavy metal and punk, and I am queer, so I learned makeup that honored those parts of my personality—and I learned it from people who shared some of the same interests. My list of beauty influences was once simple but has since become rather incoherent. It includes everything from Patrick Nagel illustrations, Nudie suits, and cowboy movies to Teddy Boys, minimalist bloggers from 2013, *FRUiTs* magazine, Gothic Lolita, James Hetfield circa 1986, and Angelina Jolie in the 1995 film *Hackers*.

As my understanding of my gender has evolved, I've mostly stopped wearing skirts and dresses, but my makeup routine has stuck. It had never been about subtlety, or "natural" femininity, so instead of losing beauty role models when I realized I was not a woman, I could look to the androgynous and effeminate men who were becoming more common among the creators of online makeup tutorials. Makeup is traditionally associated with the gender that was assigned to me at birth, but it has ultimately helped me make peace with that assignment having been incorrect.

I have always believed in being something like a logo for myself—a visual signifier of what I'm all about. My daily work uniform and my standard face minimize the mismatch between how I may be seen and who I think I am, and they maximize the confidence that I need to honestly perform my interior self. When I construct a version of myself that I like to see in the mirror, that person contains traces of aspiration that surpass aesthetics.

There is a deep sense of gender in-betweenness inside of me, and wearing a big cat eye with a boyish outfit often broadcasts that in a way that words cannot. In some spaces this may put me in danger, but for the most part it makes me feel like I have the power to make people pause before carelessly classifying me at first sight.

Author, scholar, and femme Raechel Anne Jolie writes about style and beauty as sites where rebellion, authenticity, and danger meet: "Too often contemporary critiques of beauty culture leave this key piece out—the violence of style. Not only for those who don't conform to what is acceptable, but the self-annihilation that can occur if we ignore our drive to utilize our all at once meager and potent outlets for resistance. Which is to say: if you are a marginalized person, every path comes with a cost, and often we choose the one that allows us to feel the most interior sense of livability. It is, sometimes, the only thing over which we have any control."[7]

That need for control resonates with me emotionally, and the physicist in me is keen on the word "interior." It makes me think about the interiors of strange metals, of odd ceramics that conduct electricity, or of thin layers of carbon atoms that act as if they're next to a magnet even when they're not. We don't always know what has happened in their interiors to make them appear so unusual and confusing on the outside. The way we most often try to figure this out is by putting them through a series of collisions.

When people look at me, what do I want them to see? Certainly, I am serving up a character, someone whose harsh lines seem right out of a graphic novel; someone who is so put together, so deliberately constructed that you could not possibly get to them, or mess with their self-worth. But I am also offering signs as to who I am underneath: a selective peek into my internal structure to ward off any interactions that feel more perilous, that feel more likely to make it past my outer layer and burn me in the process.

Before electron microscopes, we had optical microscopes, which were descended from magnifying glasses and the original imaging instrument—our eyes.

This is how the eye works: Rays of light enter through the cornea, then fall onto the lens, and the two parts work together to focus them. They diffract, or bend, the light rays so that they bunch up into an increasingly narrow beam that eventually becomes a single point. This point falls onto special photodetector cells at the back of the eye that then convert the light into electric signals. Those signals travel to the brain, which produces what we experience as images.

If the shape of the lens is somehow aberrant, it may cause the light rays to converge too far from the photodetector cells, and you end up either farsighted or nearsighted. Contact lenses and eyeglasses work because they help bend the light rays to focus them on the photodetectors. During my PhD training, I sometimes taught students who were preparing for medical school, and we would draw light ray diagrams on top of rudimentary schematics of an eye, then calculate the curvature and focal length of the corrective lenses that the hypothetical eye needed.

Drawing a light ray is as simple as drawing a straight line, but the concept has proven very useful to physicists over the years. At the same time, it does not capture the full nature of light. Classically, light is a three-dimensional electromagnetic wave. It is a wave of radiation, which can be visible, infrared, microwave, and so on. Physicist James Clerk Maxwell established this theory in the nineteenth century. It was complicated in the early 1900s, when Albert Einstein, while interpreting a set of experiments with light and electrons, concluded that light actually consists of particles called photons.

Today, in the wake of quantum theory, we know that light is somehow both a particle and a wave. In some experiments, you can think of it as a particle moving along a line. When it comes to others, you have to be more careful about what you imagine. What does this multitude of light's identities mean for how we see?

Whether the human eye can see single photons, particles that have no mass and no definitive size or shape, is controversial. In 2012, a team of researchers in Singapore extracted individual photoreceptor cells from adult frogs and tested them in the lab.[8] They showed that each cell could detect single photons, but there was no guarantee that this could actually happen inside a living animal.

Four years later, biophysicists at the University of Vienna ran a study with human volunteers and concluded that "humans can detect a single-photon incident on their eye with a probability significantly above chance."[9] Their result was contested on the basis of poor statistics—the details regarding what "significantly above chance" actually means—and the work to replicate these experiments more decisively is ongoing.[10]

If the human eye can perceive a single photon, that would mean that we can quite literally see a denizen of the quantum realm, which is otherwise inaccessible to our senses. Some physicists theorize that sending photons with unusual quantum properties into the eye and recording what happens could reveal more about our place in the quantum world. Humans are thought to be too big to be affected by quantum phenomena, but what if a single photon hitting a photoreceptor is a quantum process that our brain has to make sense of all the time? This sort of question keeps physicists up at night. But you don't have to go quantum to get a little disoriented when thinking about how exactly we come to see something and what exactly we are able to see.

We perceive objects, not photons or light rays, yet light has to enter our eyes to tell us that the objects are there. The brain reverse-engineers each object we see from the properties of the light that photoreceptor cells are exposed to. The mind is, in some sense, serving us a movie of the object instead of the object itself. And it is always editing that movie to make it continuous, remove tiny eye motions that would make the object blurry, and correct for the fact that the eye's lens would actually produce upside-down images if you used it to project something directly onto a screen.

These ideas were not intuitive to early thinkers. Starting in ancient Greece, scientists and philosophers thought that we can see because our eyes emit particles, rather than passively waiting for particles to hit them. Circa 360 BCE, Plato wrote in *Timaeus*, "From the eyes issues forth a stream of clear and subtle fire, of the same substance as the sunlight in the air; with which it mingles, and the two combined meet the fire proceeding from the object which is in the line of vision."[11] Although the Arab mathematician Ibn al-Haytham developed a theory that was more in line with our current understanding, Plato's idea persisted up to the seventeenth century in the West.[12]

When I was a high school teacher, I often discussed this with my students. If they could time travel, could they come up with an experiment that would convince Plato that he was incorrect? I wanted to give students a fun way to think about how they themselves come to know the world. But someone would always get stuck. "What do you mean I only ever see photons?" they'd ask incredulously, inviting me to think about the oddness of it too.

Reluctantly, I developed a bit of sympathy for all those past thinkers who'd gotten it wrong. There was something satisfying about their belief that we could be active participants in seeing, spewing fire from our eyes instead of enduring a constant cannonade of photons. And I have never been a stranger to the feeling of someone's disapproving or perplexed gaze making the back of my head feel hot.

These days, the photons that allow me to see people are often originating from a screen. When I think of beauty and style, I tend to reach for my phone, taking in TikTok recaps of couture shows, X commentary on what makes a suit look good, and makeup tutorials and "outfits of the day" across the various platforms. I have ethical qualms with fast fashion, I recognize that who gets deemed stylish often breaks down along the lines of race and body size, and I realize that a lot of what looks good on camera is due to money, lighting, and, sometimes, surgical intervention. And, because I was a subculture kid in an era before the Internet, my knee-jerk reaction to something being popular is still a little contrarian. And yet, not only do I consume this content; I have produced some myself.

I post the same kind of video on TikTok almost daily: a full-body outfit check followed by a close-up of my face and a wink at the viewer. I am not popular on TikTok, so most of these videos—almost always of my work uniform—top out at a few hundred views and, at best, a dozen likes. I make them mostly for myself, as a way of documenting who I am and who I am becoming, but the comments are sometimes rewarding too. Here are some of them:

> "Omg! Everything about this screams Betty Boop origin story 🖤🖤"
>
> "There's something Molly Ringwald about this look in the best way"
>
> "It's giving casual David Byrne. Glorious"
>
> "Nara Smith core"
>
> "It's giving Rosie the Riveter"
>
> "Holy shit. This is like wearable Chappel Roan."
>
> "I love how this sits at the intersection of Gonzo and Ms. Frizzle"
>
> "It's giving Elizabeth Perkins circa 1988 and I am so here for it"
>
> "It's giving androgyny, it's giving office siren, it's giving cottagecore corporate casual, I love"
>
> "You look like Audrey Hepburn"
>
> "I thought u were James Charles. U look great babe 🙂"[13]

I love a pop culture reference, and when people cite their sources, so these comments are like catnip for me. At the same time, I am keenly aware of how differently I would feel, and how much higher the stakes would be, if my account were actually popular, or if one of my videos went viral. Negative and offensive comments would appear more often, my privacy would become more tenuous, and with every new video, I would worry about appeasing the algorithm that had made the last one successful.

There is, after all, a whole industry that can turn a social media presence into a lucrative business, as long as your posts can be digested during a quick scroll—and accommodate sponsored content. In 2023, the business intelligence company Morning Consult found that 57 percent of Gen Z would become influencers if they had the opportunity; among all adults, that figure was 41 percent. And since the influencer world is populated with marketable stock types and personalities, newcomers can quickly learn to emulate them.[14]

While online overexposure can be a real risk for a regular person, embracing and even performing vulnerability is part of the influencer canon. You've seen these videos—they're the ones where an influencer emphasizes that social media is a highlight reel, that none of us are seeing their real problems, that we don't know how hard they have it because they are curating their life for their audience. Then, they show their tears, or maybe an anxiety-induced rash, or their "messy" apartment, which often does not look all that messy. These videos garner sympathy, they humanize the person, and they're hard to ignore. But as my best friend once noted when I was feeling guilty for not having more empathy for a crying Instagrammer, there is crying because you are sad, and then there is crying because you are sad but still had the thought to set up the camera.

In 2024, the writer Emily Sundberg offered a sharper critique:

> It's like there's this internet language that "works" for engagement (literal language, but also sense of style, and a range of trending topics to touch upon) but it all coagulates together and creates a whitewashed, boring internet. A friend pointed out that even these peoples' bad days look the same—it's never "I thought about killing my boss," or "My group dinner the other night made me super anxious but I posted it on Instagram anyway," it's always like, "wallowing, languishing, reading by the lake, journaling, feeling blue by the window."[15]

So, existing on the Internet is complex and full of seemingly contradictory options. The Internet can teach you how to build yourself an outer layer that is penetrable only to those with the right knowledge, it can put you in danger of being judged based on the quickest and shallowest photon-on-skin collisions, and it can give you a chance to monetize what is barely an imitation of exposing your inner layers.

I first met many of my close friends on social media, and a lot of my romantic relationships also started online. This is a symptom of both how much I like having control over my presentation and how relatively easy it is to find people online who share my interests. It's also true that social norms have shifted away from meeting people in other ways, and I am not enough of a contrarian to resist that. But although Instagram friendships and message-based "flirtationships" may be easier to come by, they also have a worrisome feature: when they move into physical spaces, much more becomes much harder to control.

There is the issue of awkward silences, of rapport and chemistry, of making eye contact too much or not enough, or just plainly not liking someone when they are not flattened into a screen. More mundanely, I also worry about not looking like my pictures. Do I look less optimal when viewed from angles other than the ones I favor for my #ootd (outfit of the day) posts? Does my eyeliner look less sharp up close? Is my skin splotchier under the orange lights of coffee shops and restaurants? Am I taller than they expected? Slouchier? Wider?

This kind of thinking is unfair to my companions, and it is self-centered, but I fall into it anyway. The photons that must hit me to help someone else notice the color of my hair or the angle of my jaw also leave me in a state of slight emotional recoil, a simmering worry over what will happen to the information they have gathered about me from that collision.

Will the light hit me at all the wrong angles and produce a distorted, untrue version of me? Will it represent me perfectly but elicit disinterest or disdain anyway? Or will the light somehow reveal more than I think it's capable of showing—in which case my metaphor will fall apart and my new friend will learn too much about me too quickly?

I learned these fears from the story of the electron microscope: sometimes being seen too deeply can leave you burned. And I learned them from my father, who always had a gift for what you might consider interpersonal, emotional spectroscopy. He has always been able to determine a person's most interior pain points simply by observing the thoughts and energies they emanate into the world. In our moments of conflict, he has gone straight for those points. Growing up, exposing my inner self always felt like a risk, and everyone in my family seemed to know that, when going into a tense situation, you do not bring a tool for gathering more information—you always bring a weapon.

Consequently, as a young adult I was riddled with self-doubt. In high school and college, people who tried to intertwine their lives with mine found that beneath my well-curated outer layer and no small amount of cool brashness, there was mostly self-consciousness and fear. If they tried to probe deeper, they would trigger a dramatic reaction, waves of suspicion and anger, and then, often, tears. Either I was in control, or in absolute shambles. Like my parents, I never learned to say, "Can we take another look at what you mean?" or "Can you please show me more of your thinking?" I just felt the metaphorical particle hit me and reacted instinctively.

But you can only be so reactive for so long before it exhausts you. One solution that helps preserve energy, and which was very much in line with the form of irony popular in the 2010s, is to let every kind of inquiry, observation, and criticism bounce off you. When I became tired of actively obscuring my interior, the cultural script of a detached coolness was waiting for me.

I aspired to act and look like an incredibly calm and collected character from a song, or a movie, or a dream—one whose face was always painted to signal that they just couldn't care less. That person is still with me, in a lot of those TikTok videos. Their sharp eyeliner and high, highlighted cheekbones are in part there to let you know that this person is too calm to argue, too self-assured to be hurt.

But that person does not really exist in my close friends' experience of me. That may be the person they first met, but not the person who took part in establishing the closeness that we now share. I realized this a few years into my thirties while I was telling my best friend about a series of failed attempts to start an off-line relationship with an Instagram mutual. In addition to being my best friend, they were well positioned to advise me because we had also met on Instagram years ago.

"It really sucks that they are missing out on you," my best friend said. I protested, saying that I am really easy to get to know because I share so much of myself online. They disagreed: "This person just has an idea of you, this cold, determined scientist person that they can't see past." I probably don't have to tell you that time proved them right.

I often think about how they are one of the few people that has ever called me goofy. Another thing they said stands out in my memory: "You have this face you sometimes make, it's like, 'I'm just a guy.' It's so endearing." I was floored. For all the time I invest in putting on sharp, polished faces every morning—eyeliner pen, brow gel, and lip liner firmly in hand—this was the face that had stuck with them. And it did not hurt that they could see me so clearly—it only made me feel loved.

One of my favorite Instagram meme accounts is called @softcore_trauma, run by the writer Margeaux Feldman. Their posts feature images of Muppets, cats, and other cute animals overlaid with text on topics like trauma, healing, queerness, and intimacy.[16] Feldman is a gifted collage artist and a scholar of literature and psychology, so their page is both fun and emotionally resonant.

In one of their memes, an image of a rose on a pillow is superimposed with the text, "When you trigger me, I'm being given an opportunity to turn my attention towards the wounds that need healing."[17] In an essay that accompanies the Instagram post, Feldman, expanding on arguments made by queer theorists Lauren Berlant and Judith Butler, elaborates: "What if we stopped seeing being triggered by those we love as a negative thing, but rather as a wound-opening/world-opening portal to a love that is as rich as it is inconvenient? This is the kind of love that I want."[18]

I wondered about the worlds that I could open in this way. Could it be worth the gamble of getting hurt when I let someone see past my style and into my interior—the messier, uglier, and weirder version of myself?

I once loved a woman who sometimes wore makeup but was never not suspicious of it. She owned the absolute minimum of products, and she consistently argued that washing your face with anything but plain soap was equivalent to getting scammed by the beauty industry. I gifted her vibrant eye-shadow palettes, talked about the virtues of moisturizing, and took her lipstick shopping. Sometimes, I was amused by the queerness of it—by me, the transmasculine person in pants, pleading with my cisgendered girlfriend to recognize the importance of properly applying concealer. But some of our makeup arguments were too loaded with subtexts of politics and identity to be amusing.

We'd debate whether it was possible to dislike makeup without that dislike sitting awfully close to internalized misogyny. Whether it was possible to wear makeup without unwittingly supporting the patriarchy. Whether bold makeup was just as much a defense mechanism as makeup that looked more traditionally feminine—and therefore was not bold at all. We argued about all of this and more, never fully agreeing on how a person should live their life for the good.

She challenged me on being too rigid in my assessment of people, and she reminded me over and over to be more curious and patient, to find techniques for seeing others better, to calibrate how I probed her opinions and analyzed her behaviors in order to foster discovery rather than cause harm. She taught me that even bad arguments don't have to be the end of a relationship—that damage can be repaired as long as there is a will to do so. And, at her best, she understood that for us to build a love and a life together, we had to make ourselves visible to each other beyond our outermost layers. The moments when she would say "Thank you for seeing me" remain some of the most tender I have ever experienced.

One spring, while we were still dating, she told me she'd been trying to figure out cat eyeliner for some occasion, and I took it upon myself to make her a helpful video. I'm awkward in it, trying to come off like a TikTok makeup influencer but with zero production value and no editing skills. Yet, months later she would show up with the eye executed perfectly, the sharp tip as deadly as ever, leaving me temporarily breathless.

CHAPTER 10

WAVE-PARTICLE DUALITY

The word "queer" shares a common etymology with "torque," which is a term that I learned in physics classes. While a force pushes you to accelerate or decelerate along a straight line, a torque will make you twist and rotate. We call this "angular" motion, because the angle between you and the point around which you are rotating changes. Nothing has made me look at things more closely—at myself and my world, from every angle—than the realization that I am queer.

ME AND MR. DAVISSON

Before I tell you more about myself, I want to tell you about Clinton Davisson. He was an American physicist, and in 1926 he was in England, trying to enjoy a "second honeymoon" with his wife, Charlotte.[1] At some point, he ducked out to attend a physics conference and was surprised to learn that data from one of his experiments back in New York was being used to prop up a radical new piece of physics called quantum theory. Among other things, it posited that electrons, even though they were particles, could behave as waves. This idea was first put forward in 1924 by Louis de Broglie in his PhD thesis, but by all accounts, Davisson hadn't read it before quantum physics intruded into his vacation.

Davisson's experiment had been relatively simple. Inside of a tube that had been emptied of all air—a vacuum chamber—a piece of the metal tungsten was heated until it released electrons. These electrons were then exposed to an electric voltage that accelerated them toward the other end of the tube. There, they encountered a chunk of nickel, collided with it, and scattered toward a detector.

Davisson and his colleague Lester Germer had been studying this process because it was somewhat related to a certain new type of amplifier that their employer, the telecom company AT&T, was using in its transcontinental telephone lines. They hadn't really set out to change our fundamental understanding of electrons. They inadvertently built an experiment that would prove able to detect wavelike behavior in the electrons, even though such a possibility had not originally occurred to them. The atoms in the nickel were arranged into neat lines separated by neat gaps so that when an electron encountered it, there was a measurable difference between it crashing through like a wave and it bouncing off like a particle—and this difference would show up in the detector data.

When Davisson left for England, the experiment did not seem to be going well. The two men and their colleagues couldn't really explain the data from the detector. But Davisson and his colleague Charles H. Kunsman had published a paper about a similar experiment with platinum and magnesium in 1923, three years earlier, and some of the physicists he spoke to in England had read it through a quantum lens.[2] Speaking with them put the wave idea in Davisson's head, and he returned to the lab with a new purpose. Eleven years later he, along with George Paget Thomson, was awarded the Nobel Prize in Physics. The title of his Nobel lecture was "The Discovery of Electron Waves."

I learned about the Davisson-Germer experiment in my quantum physics classes in college at the University of Chicago, where Davisson had also studied. I'd chosen to attend this school because the physics department was a storied one: over the years, paradigm-shifting researchers like Enrico Fermi, Albert Michelson, and Robert Millikan had done their work there. Davisson had rubbed shoulders with some of them. He'd been especially inspired by Millikan, another big player in the history of the electron, who'd made Davisson "delighted to find that physics was the concise, orderly science [he] had imagined it to be."[3] Davisson and I also have this in common—the promise of putting the world's ingredients into precise order has always appealed to me too.

Physicists ask big questions—like why there is something instead of nothing in the universe, and how it all started—but they can only tackle them by breaking them up into smaller questions first. One of the strengths of physics is exactly that it gives us recipes for systematically disassembling the world into building blocks and capturing their functions with symbols and equations. To do physics, you often have to first sort the ingredients of reality into categories. For instance, if you handed me a particle, I could tell you whether it falls into the category of "fermion" or "boson," and its category would indicate whether it carries a force or makes matter. It can be emotionally satisfying to know what a thing is on the basis of just one word.

I'd gravitated toward this framework as a child when I badgered my father with questions like "What is electricity?" instead of trying to learn what it does. Later, like a true future physicist, I explained to my younger brother that electricity can be broken down into electrons. As a young physics student, I thought that the key to understanding anything physical was just knowing enough math. If you were good at math, I reasoned, you could determine what category a thing belonged to, then find equations describing that category's behavior, then solve them to predict the thing's future.

This was, in a sense, the promise of classical physics as it was practiced before the 1900s. This was also the physics I had learned from seventh grade up until my second year of college. Such an approach, from categorizing to calculating, seemed to me to chart out a path toward knowledge that was both sensical and plausible. When I learned about quantum theory, then, it should have rocked me to my core. Instead, surprisingly, I fell in love with it.

Quantum theory still lets you break the world into pieces, and it has solid mathematical bones. You still do a lot of sitting down and calculating, just with different equations. But making sense of those calculations becomes challenging and, often, shocking. Determinism and neat, mutually exclusive categories fall away. Fuzzy probabilistic predictions and paradoxical conclusions abound.

In the 1920s and '30s, physicists talked about quantum behaviors as "spooky" and "burlesque." Contemporary physicists—and science journalists like me—now favor a descriptor like "counterintuitive." One of my favorite examples is a 2021 *New Scientist* article in which the physicist Carlo Rovelli states that "quantum weirdness isn't weird—if we accept objects don't exist," a trade-off that most people would never consider viable.[4] Clinton Davisson was aware of the weirdness of quantum theory, too. In his Nobel lecture, he referred to electrons as "particles which were ignorant apparently of even the rudiments of dynamics," and characterized some of the data on their behavior as a "perversity."[5]

This is where I have to disagree with Davisson. Quantum theory never upset me, and though I was repeatedly, in numerous classes, told that it was odd, I never really had that impression myself. Maybe it is because I first encountered it in a popular science book when I was fifteen and was too young to fully grasp its implications. Maybe it is because, as much as I was drawn to ordered, systematic, well-delineated analysis, there had also always been something "perverse" about me. Maybe it is because I have always been queer.

The historian Beans Velocci writes about how, as a nonbinary person, they are drawn to classifications in science—they feel torn between wanting to fully understand these systems that define normalcy (which they find themselves standing outside of) and wanting to disprove them. Velocci recounts: "Certainly, my own non-fitting into one of two invented-but-naturalized sex/gender categories made me particularly prone to analyses of classification. . . . But with the encouragement and tools of this literature, I began to look for it everywhere: medical diagnoses, species distinctions, what counts as a soup."[6]

Though the mention of soup is meant to be humorous, I have in fact considered all three of Velocci's examples. I have read books about why "fish" is such an ill-defined category that it might as well not exist, and I have had plenty of playful arguments about what the proper definition of soup is, as well as what makes a sandwich a sandwich. My interest in categories has extended beyond my interest in physics, and I seem to always be bumping into those without strong edges.

I embraced the language of nonbinary gender as an adult, after years of being less comfortable with my body than with the intricacies of the universe. The prospect of what accepting my honest self might do to my reality scared me more than bold physics theories that framed our entire physical reality as a hologram or simulation. But even at fifteen, when I thought and spoke of myself differently, I could accept that an electron can be particle-like without fully being a particle, and wavelike without fully being a wave—that it was somehow simultaneously neither and both.

Velocci again: "I am not saying that every trans or nonbinary person is going to share this suspicion of neatly delineated categories—which should come as no surprise, after all, given that very suspicion. But for me, at least, through the bringing together of histories of sex science and theories of classification, there was a distinct gestalt shift that happened, from *sex isn't real* to *probably nothing else is either*."[7] Physics played a similar role for me. I gravitated toward topics that would expose me to instances of categories being broken, maybe even dissolved. Similar to how Velocci uses "nothing else" as broadly as the reader wants to interpret it, discovering quantum theory made me rethink the role of assigning a fixed identity to, well, anything at the scale of the whole universe.

And I had been truly enamored with quantum theory from the jump. I incessantly asked my high school teachers about it, tried to recount whole books about it to my mom, and counted down the days until I could rigorously learn it in college. My PhD focused on quantum phenomena, and writing about all things quantum has been the backbone of my career in science journalism. In the newsroom where I work, I can't seem to stop getting into heady discussions about the quantum nature of our world.

"'Chances are the whole world is quantum' is very upsetting to me and my conception of reality," my editor recently wrote to me, quoting my own assertion.[8]

I reacted with a smiling emoji, and later wrote about the study that had prompted this exchange. But what I really wanted to say was, "You don't get it. The quantum world was the first place where I knew queerness, even if it wasn't going by that name. Meeting its not-wave-not-particle denizens was the first time I ever saw something like a possibility of myself. How could you not love that reality?"

THE BIOGRAPHY OF THE ELECTRON

To understand why the Davisson-Germer experiment was so important for quantum theory, it's worth revisiting the biography of the electron.

Its story usually starts with the experiments of Joseph John Thomson in 1897, but the idea that the electrical charge was a particle preceded his work by more than a century. There was Benjamin Franklin in 1750, expounding that "the electrical matter consists of particles extremely subtle." In the 1840s, the British natural philosopher Richard Laming conjectured about "the existence of subatomic, unit-charged particles." James Clerk Maxwell was discussing the "molecule of electricity" in 1873, and Hermann von Helmholtz was writing about electricity being divided "into elementary portions which behave like atoms of electricity" in 1881.[9]

Many of these researchers were basing their ideas on contemporaneous chemistry experiments, on empirical realities that hinted at the existence of a particle-like entity that was making certain reactions happen. In his own biography of the electron, the historian Theodore Arabatzis explains that because researchers of this era understood that all matter is made from atoms, they were primed to think that something like electricity could also have smaller constituents, and that they would be particles.[10]

Joseph John Thomson's experiments involved electrons moving through a vacuum tube in the form of a beam, or cathode ray, that produced a visible glow when it hit one of the tube's ends. Because electrons had not officially been discovered yet, there was an ongoing debate on what exactly cathode rays are, and Thomson hoped that he would find out through these experiments. He discovered that magnetic and electric fields could bend cathode rays, and theorists quickly ascertained that this could only happen if the rays were made from particles. Just like that, Thomson discovered the particle that we now know as the electron.

He determined how strong a field had to be to bend a ray by some specified amount, then used that information to calculate the ratio of the "corpuscle's" mass to its electric charge and also estimate its size. "The mass of the corpuscle is only about 1/1,700 of that of the hydrogen atom," he wrote, thus unseating this atom as the paragon of smallness.[11]

Thomson found that the corpuscle always maintained the same behavior and mass-to-charge ratio, even when the conditions of the experiment were switched up. "Wherever it is found, it preserves its individuality," he wrote, conjuring the image of the electron as something distinct, with discernible edges.[12] Ten years later, Robert Millikan, who was the first to measure the charge of the electron, speculated that only a completely new kind of physics could allow us to break the electron into something even smaller.

So far, he has been correct. The electron is an entry in the Standard Model of particle physics, and theorists broadly agree that there is nothing smaller hiding inside it. But history was less kind to the idea of electrons being corpuscles, especially after Davisson and Germer.

When Davisson and Germer fired electrons at nickel, the electrons did not behave as individual particles at all. Instead, they acted much more like waves of water crashing into the shore. Shortly after, in an independent set of experiments, George Paget Thomson—the son of Joseph John—shot electrons through thin metallic crystals and observed the same wavelike behavior, contradicting the celebrated work of his father. None of the men who had been writing the electron's story for the past century could have imagined this.

In classical physics, waves and particles are fundamentally distinct. Particles are localized and finite bits of something, while waves are disturbances on some medium and extend as far as that medium does. A droplet of water has edges, and you can convey its size with a number. But a wave only stops when the ocean does. A particle then may be part of a wave, but never identical to one. Or at least that's what physicists thought until de Broglie's radical 1924 idea.

De Broglie went as far as claiming that all particles may also be waves, but his case for the waviness of the electron was particularly compelling. His calculations suggested that if electrons were not wavy in some way, atoms would not be stable, and the physical world would crumble. Imagine an atom as a tiny solar system, its electrons orbiting the nucleus like planets around the sun. This is an imperfect picture that does not fully convey our best quantum mechanical understanding of the atom, but it helps illustrate the gist of de Broglie's argument. Specifically, resolving the question of how electrons find stable orbits is crucial for explaining why atoms are stable. De Broglie calculated that, for wavelike electrons, stable orbits corresponded to standing waves, like the sustained vibration of a violin string. This was an elegant idea, and it fit right into the theory of atoms that physicists like Niels Bohr were developing. But no one had been able to observe electrons inside of atoms to confirm these ideas. Before Davisson, Germer, and Thomson, no one had caught them being wavelike either. Davisson and Germer didn't even seem to know that someone had suggested this was an option.

Thomson had known of de Broglie's work and was interested in it, but the two other men had not set out to question the classification of electrons as particles. They certainly did not mean to open the door to question the character of all matter. But that was an inevitable effect of their work. All matter is made of atoms, and all atoms contain electrons. So, there is something wavy about every single thing you have ever touched, including yourself. But it often takes a lot of effort to see that.

MORE THAN AN ALLY

I started publicly calling myself nonbinary in 2020, when I was a high school teacher. I added "she/they" to my email signature and nervously told students that they could refer to me as either. At first, the idea of audibly saying "I am nonbinary" felt dangerous and destabilizing, but whenever someone referred to me as "they," I felt a new version of myself emerging. Before introducing that person into my classroom, they had been just skulking at the edges of my thoughts. I liked them more now. I mean to say: I liked *myself* more.

I had met nonbinary people before 2020, in college and in graduate school, and I had encountered the basics of queer theory in online articles—as well as daily blogs of trans people. Queerness and transness had long been familiar ideas to me. In graduate school, I'd organized with the "women in physics" group and had supported renaming it to "women and gender minorities in physics." Back then, I probably included my pronouns in introductions more often than I do now, always so vigilant to make space for trans people in my communities. I'd never really understood how exactly one is supposed to be a woman, but for years I only ever thought of myself as an ally to my trans kin. That was the only thought that, at the time, felt safe.

I had long been drawn to the blurring of boundaries between categories, and the excitement of existing in indeterminate middles, but I questioned whether I could belong there. At one graduate school event, I told a younger queer physicist that I had "an academic interest in being a little more butch," then had to twist myself into verbal knots to explain what that meant. This was how I used to think about my gender—I intellectualized it to avoid feeling and living it.

To dull some of the discomfort of existing in my body, I tried on different kinds of womanhood. Feminism had taught me that there is no one way to be a woman, that women came in all sorts of shapes and sizes and flavors, so I kept trying to find mine. I never did. Instead, all of it—from being a hot goth girl to a '50s vamp to embracing something like androgyny—still felt like cosplay, fundamentally unlike what I saw in "real women" around me.

What changed in 2020 was that I was finally living with my partner with whom I'd been maintaining a long-distance relationship for nearly six years. We had both left our isolated graduate school towns to be New Yorkers. The COVID-19 pandemic was wreaking its devastation, but in its wake I had the chance to look for community in a place that was bigger and freer than any I had called home before. I was rebuilding my life with the one person that always made me feel safe and understood. Tellingly, the new people I connected with the most were queer, and many were also nonbinary.

They were the grown-up versions of gender renegades I had met in more niche spaces at universities. They had jobs, pets, partners, and hobbies. Their lives were queer, just like they were, and even when those lives were hard there was still room for queer joy. They had not grown out of their gender identity crises—like the one that I had begun to ascribe to myself in a joking manner before I was able to confront it more seriously—but had resolved them through acceptance. I'd been so caught up in pursuing an academic career and in projecting the image of a green card–worthy immigrant that I had not considered the possibility of building a life like this. Now, however, the conditions were right for me and these new, bright people in my life to "midwife for one another the image we form of ourselves," as the writer Raquel Gutiérrez puts it.[13] My friends did not make me nonbinary, but they did make it feel safe for me to accept that I am.

And then there were my queer students. Though only a few were trans, they had an immense impact on me. If a fifteen-year-old could advocate for their "they" pronoun, or for being called by their correct name in front of teachers who were known to be transphobic, what was I—an adult with a job and tons of privilege—afraid of? And did those kids not deserve to see that someone like them could make it from the back of the classroom to its lectern? I was teaching them how to use physics to understand the world, but they inadvertently taught me that I shouldn't be afraid of understanding myself.

In 2024, I returned to this school to speak about my work in science journalism, and I caught up with some of the students who had been freshmen when I taught there. Now they were about to graduate and leave for college, and a few stunned me by saying that they were considering majoring in physics because of our classes together.

"I wrote my college essay about this one queer physics teacher who once told me I could do physics too," one student said. The word "queer" felt so deliberate. It meant so much to hear it. When I'd been their age, especially while I lived in Croatia, all I'd had was quantum theory's suggestion that it's okay to be neither a particle nor a wave.

EXPERIMENTAL CONTRADICTIONS

At the core of what makes the wave-particle duality upsetting to scientists is the way it pits experiments against each other. If you believe, as many scientists do, that the world has certain physical properties and that the task is simply to find the best way to observe and measure them—that all questions can be answered by devising cleverer and more precise experiments—then the prospect of two experiments contradicting each other must mean that one is somehow defective. With time and hard work, you should be able to determine which one is correct. But for electrons, the opposite has been the case.

After Davisson, Germer, and Thomson, many research teams subjected electrons—which still behaved like particles most of the time—to tests of waviness, and the electrons always passed. In 1961, for instance, the physicist Claus Jönsson put electrons through the double-slit experiment that had historically been used to prove that something was a wave.[14] Here, electrons traveled toward a wall with two very thin slits, passed through them, then hit a screen. A true particle can go through one slit only, and if you repeat the experiment many times, you should end up with roughly the same number of marks on the screen right behind each slit. A wave, however, can pass through both slits at once, burst out as a wave of a different shape, and leave a distinct stripe pattern on the screen. Jönsson showed that electrons recreate this stripe pattern, providing such a definitive example of their wave nature that it is commonly featured in textbooks.

But the electron was actually not physicists' first brush with the wave-particle paradox. In the 1900s, they had been faced with an example of how the trouble with labels can go the other way—how something that ought to have been a wave can sometimes also behave like a particle. This troublemaker was light.

EMPIRICAL IDENTITY JUDGMENT

Unlike the electron, I thought I had some say in when, and to whom, I would reveal my double character. I'd known I was bisexual around the time I started high school but kept finding reasons to not come out. Most of them were based in fear. First, fear of judgment from my parents, then from my peers, then, later, from administrators at my expensive American boarding school. In college, I was afraid of not being accepted by queers who were already proudly out of the closet, and of being fetishized by straight people who thought queerness was edgy. Always, I was afraid of physical violence. But as I grew older, I grew more resilient, and more tired of secrets. In the absence of me making any grand announcements, my friends started to understand that I wasn't straight. When I went to graduate school I considered myself "out," but still tried to be very nonchalant about it.

When I did speak up about it, such as on social media on National Coming Out Day, for the most part no one cared. I was both relieved and disappointed, and always frustrated. Years later, my mother-in-law inadvertently explained to me why people generally didn't seem to care: by marrying her son, she said, I had effectively "chosen a side," and that was all that mattered. In other words, people felt that, unless I had a girlfriend immediately in hand, they could disregard my stated identity. Once I married a man, my mother endorsed a similar view. During a visit, she conjectured that I must be having an affair with a female friend, because otherwise I would no longer be bisexual. I had read about "bisexual erasure," and I understand that being coupled with a man makes my life materially easier within the still abundantly straight world we live in, but these conversations still made me feel alienated and lonely.

People often judged my identity based on what they could observe. Seeing me with my male spouse counted as convincing experimental evidence of my presumed heterosexuality. In the experimental setup established by the unstated question, "Will Karmela hold his hand?" my sexuality is cleanly delineated the moment I do. My mother's idea actually had a little more in common with fuzzier quantum situations. She acknowledged that another experiment could be cooked up—the affair—to reveal that I had another nature too. The idea was offensive, feeding into the stereotype of bisexuals as insatiable cheaters, but I couldn't help but sort of appreciate it.

Other people similarly perceived my gender identity, including my coworkers at the high school. Unlike my students, my coworkers never called me "they," and they remained largely uninterested in how I saw myself. I tried to put myself in their shoes: it would have taken conscious effort for them to revise their conception of me, and as pandemic teachers we were all exhausted. I didn't want to be another complication to anyone's work.

On my first or second day, a colleague asked me and another new hire whether we "had pronouns." The tone of the question made my heartbeat double. I lied, then instantly regretted it. I made no lasting friendships while working there.

It didn't help that 2020 was the beginning of people taking a more overt political stance against the existence of trans people—one that continues to endanger us, especially the people of color in our community. Each year since has brought a new record-breaking number of anti-trans bills introduced in legislatures.[15] Conservative politicians also continue to campaign on lies about trans people, threatening their livelihoods and safety.[16] In January 2025, President Trump signed an executive order that stated it is "the policy of the United States to recognize two sexes"; it also used alarmist phrases such as "gender ideology," which directly originate from some of the most transphobic Internet circles and strands of public thought. Later in the year, the US Supreme Court upheld the legitimacy of a ban on transgender healthcare for minors, signaling to states that it is okay to prevent trans youth from becoming trans adults. Since 2020, my sense of certainty in my identity as a trans nonbinary person has been rivaled in magnitude only by fear of what law or campaign may make it dangerous for me to say that out loud.

Witnessing this rhetoric take hold among some members of my extended family scared me in the same way that homophobic comments had scared me when I was a teenager in largely Catholic Croatia. Though Croatian laws do currently offer legal protections for queer people, bias against them is still rather strong. Croatian politics has moved rightward in recent years, and homophobia and transphobia have only become more pronounced. The prospect of discussing gender with my Croatian family is deeply uncomfortable, my mother being the sole exception. And it has, in recent years, been terrifying to see my American in-laws read newspapers that demonized people like me, and vote for political candidates who promise to erase us. How could I ever come out to them as nonbinary? Feeling cowardly, I stayed in the closet even when I lived in their home.

When I went to work in a newsroom, I was less naive about how I would be perceived. It still took time to find the courage to insist on only being called "they," but I now knew that I would have to keep coming out over and over in any scenario. And I knew to brace myself for possibly being met with disapproval and hostility. I got lucky in that this never went beyond colleagues quietly endorsing the work of transphobic writers or exploiting ambiguities in our publication's style guide to excise gender-neutral language from their work. The vast majority of people I work with day-to-day have been kind and supportive, and their support has only become more meaningful to me over time. I doubt that any of those whose actions made me feel otherwise ever made the connection between these small acts of erasure and me, their actual colleague. I'm sure they had no idea that I noticed. I'm certain they still see me as only what they want to see—as the thing that they have built their observation apparatus for.

It is also a sign of my privilege—as a white person, as a person who can still pass as cisgendered, and as someone who had access to higher education—that I have a stable job at all. It was only as recently as 2020 that the Supreme Court ruled that employment discrimination against transgendered people was unconstitutional.[17] And without steady employment, people end up at risk of housing insecurity. Survey reports vary, but the proportion of trans people who are experiencing houselessness ranges from 35 percent to as much as 63 percent.[18] Writing this from my one-and-a-half-bedroom New York City apartment, I am so much safer than many of my trans kin will ever be. Because my transition so far has not included medical interventions, I can travel without fear of being denied access to essential care. My material reality is objectively easier to navigate than for many in my community. But even though I acknowledge that I am so lucky to rarely have to worry about anything beyond being misgendered, no amount of gratitude for my socioeconomic position can fully dull the sting of people dismissing who I know myself to be.

"GNOME WITH TWO HEADS"

Though it was once a topic of heated debate, most particularly during the time of Isaac Newton, the nature of light seemed to have become a more settled issue by the beginning of the twentieth century. All experimental evidence pointed toward light being a wave, and physicists broadly accepted this. Unlike the electron, light went a whole century without an identity crisis. Then, Max Planck took up studying how objects radiate.

When he assumed that objects radiate light waves, the math produced an outrageous prediction—it seemed to suggest that any warm and dark object, even a cup of coffee, would radiate infinite amounts of ultraviolet light. To correct this, Planck assumed that objects emitted light in "chunks" or "quanta" instead of waves. This was an unusual idea, but what the equations suggested now matched reality. One of them, now known as Planck's law, has since helped us understand the radiation of many kinds of stars.

In 1905, Albert Einstein added credence to the idea of light being composed of particles, or photons, when he explained how shining light on a piece of metal makes electrons jump out of it. The wave theory of light explained how light was transferring energy to these jumping electrons, but it could not explain a peculiar observation from certain experiments: the particles only jumped when the incident light had specific frequencies, which correspond to specific colors. The researchers, thinking of light as a wave, had expected that if they turned any frequency of light to a sufficiently high intensity, they could get some electrons to jump out of the metal every time. Yet they found that for some frequencies, even high-intensity light proved ineffective.

Planck's photons had energies directly proportional to the frequency of the light, so they had little energy that they could pass on at low frequencies. Einstein suggested that, when light fell onto a metal, photons were colliding with electrons like billiard balls. In this scenario, low-energy photons, or low-frequency light, could not sufficiently "bump" the electrons. From this hypothesis, Einstein calculated the threshold frequencies for pushing electrons out of specific metals, such as gold—and they matched what happened in experiments.

In the aftermath of Einstein's work, physicists had to add "wave-particle paradox" to their lexicon. By the 1920s, most physicists accepted, though sometimes begrudgingly, that this way of existing was an option within our physical reality. Davisson remarked on it in his Nobel lecture, saying that "what still appears to many of us as a contradiction in terms had been proved true beyond the least possible doubt—light was at once a flight of particles and a propagation of waves," and that consequently, "light, the perfect child of physics, [had] been changed into a gnome with two heads."[19]

This may sound dramatic, but as physicist and philosopher Karen Barad puts it, "It was not merely that new empirical evidence concerning the nature of light seemed to contradict the established view, but during the first quarter of the twentieth century, it became increasingly difficult to understand how any consistent understanding of the nature of light would be possible."[20]

So, suppose that you are a young physicist in the 1930s, and you want to take stock of everything that exists in our world. Could you still argue that some objects are localized and can be called particles while some are extended and can be called waves? The most correct thing you could actually say would be that everything is somehow neither and also both.

Bohr referred to this as "complementarity," positing that being a particle and being a wave are in general not mutually exclusive states, but that you also cannot construct a single experiment that will simultaneously reveal both. And physicists did keep building and imagining experiments. In some, they found tricks for making particles revert to their particle-like nature, even in the double-slit experiment.[21] If the double-slit is modified so that you always know which slit some amount of light has gone through—for instance, by adding a detector to the slit—you get not the interference pattern characteristic of waves, but rather two marks on the screen, which are characteristic of particles. Here, obtaining extra knowledge about light seems to change its nature. Researchers also showed that in a less extreme case—where you find a way to learn just a little more about which of the slits light may be going through, but not with absolute certainty—the striped pattern on the screen does not disappear, but it does become less definitive, as if the light has lost some waviness and gained some particle-ness.[22] This is an intriguing advance, but nothing about what wave-particle duality truly means has been fully resolved yet.

"If you talk to the community, maybe 50 percent would say this is normal because it's quantum physics, and the other 50 percent would really scratch their heads because it's quantum physics," says physicist Markus

Arndt, whose own experiments show the waviness of large molecules.[23] Many physicists agree that because quantum theory does not provide more clarity on this issue, it is an imperfect theory and therefore must eventually be replaced with a better one.

There are several interpretations of the mathematics of quantum theory that sidestep ideas like complementarity, and many are still being actively developed today. I've written about some of them as a journalist and found them interesting. One research team, for instance, developed an analysis of the double-slit experiment where the striped pattern follows not from the wavelike character of light, but rather from some photons combining into special quantum states that the detector simply cannot pick up.[24] It was exciting to speak with them, but also rather difficult to find other physicists that would sign off on their idea being a real game changer. The refusal of matter to fall into a neat binary has not yet been explained well enough to convince all physicists to abandon Bohr's idea, even if that idea isn't satisfying. The consensus is still with a neither-and-both that does feel peculiar, even bothersome, if you have never felt it within yourself.

QUEER PHYSICS VS. QUEER PHYSICISTS

While the existence of my husband is enough for many straight people to deem me a cisgendered straight woman, other queer people often express incredulity at how anyone could possibly think that there is anything "cis-het" about me. Of course, I can see where they are coming from.

My short hair, my boyish style, the drag-like flair of my makeup—do they not all signal something? Can people not see queerness in the way I walk and sit? Can they not see it in the shape of my shoulders and thighs? "Nonbinary" has been such a comfortable home for me exactly because it feels like the truth of my body beyond what I choose to wear. As the author Melissa Faliveno wrote about her own androgyny, "It isn't just the choices I make about my appearance that make me androgynous but the body I was born with, the DNA that built me."[25] I do not want to endorse anything like biological essentialism, but my own unfruitful attempts to learn womanhood through style have also taught me not to ignore the tangible realities of my body.

I've always been tall and flat, I could always gain muscle, and even when I didn't work out regularly, I never really grew soft. As a long-haired teenager who wore a lot of oversized band T-shirts, I'd sometimes get mistaken for a boy, and cutting my hair only amplified that confusion. Short hair didn't necessarily make me more masculine, but it did let the sharp lines of my face come into their own. "You could have worn anything, and I'd still have clocked your queerness," an ex-girlfriend once told me. "No one does gender like Karmela does," her ex-girlfriend concurred.

"Of course you would think that," I am tempted to respond when I hear comments like these. "You are like Clinton Davisson after he returned from his second honeymoon. You know what to look for—you know what the possibilities are!"

But a joke in which I am the perverted electron or the gnome with two heads would not really have an audience to land with because queer physicists are few, many of them are still in the closet, and most don't feel like laughing. Surveys of their experience as physicists reveal high levels of anxiety, depression, social isolation, and instances of harassment. A 2016 report noted that over 40 percent of 324 surveyed queer physicists agreed with the statement, "Employees are expected to not act too gay." About 30 percent of gender-nonconforming and trans physicists described the climate of their department or division as "uncomfortable" or "very uncomfortable," and over a third of the survey respondents said they had considered leaving the field in the previous year.[26]

In a 2024 survey of a hundred women and LGBTQ+ physics doctoral students, ninety-six "reported neutral at best and wholly negative at worst experiences in their doctoral programs," with many describing those experiences as "toxic."[27] Physics departments are riddled with instances of some of our brightest minds remaining firmly closed to the possibility of respecting queer personhood.

Chanda Prescod-Weinstein, who was the first publicly queer physicist I knew of, writes about the paradoxical nature of such moments: "Somehow we can do book-length quantum field theory calculations that are predicated on . . . nonbinary understanding of particles, but we can't commit to learning a couple more words when we learn someone's name or acknowledge that our categories are failing us socially."[28]

This restricts our understanding of the universe. It limits us to the intuitions and imaginations of those who are deemed "normal." It leaves behind all the insights that those who experience the world differently may have had, had they ever been asked to reach for them. And it robs those individuals of the joy of discovery. The result, as Prescod-Weinstein says, is "an injustice that impoverishes humanity" and an injustice "to those who are denied an opportunity to walk the intellectual pathways that their brains might unfold."[29] The status quo is one where everyone loses.

"THE EDGE OF KNOWLEDGE"

The wave-particle paradox is most challenging, and most profound, when you try to think about an electron or a similar object when it is unobserved. What is the electron when no one is asking it to reveal itself? What changes once you put it in an experiment? Does the electron somehow "recognize" that it is in an experiment meant to uncover its waviness or particle-ness, then respond by acting as that wave, or that particle? Or is there a more fundamental reality of the electron that is simply inaccessible to us? Physicists have argued about this for decades, and the electron has somehow managed to avoid telling them when, or whether, they may have been right.

There are more than a billion billion billions of electrons in my body, so this not-particle-not-wave and I will never actually share the same physical experience of the world. But I find it reassuring that, at least in theory, there are instances where its identity seems to be indeterminate. When it comes to humans, although media representations sometimes flatten "nonbinary" into something like a third gender, I understand it to be a spectrum of these indeterminate moments—of just existing, not constricted by questions, unencumbered by a search for truth beyond some inner feeling that is only one's own. If this mode of being is allowed for the electron, I don't see why I cannot claim it for myself as well.

This is not to say that embracing queerness lets me off the hook intellectually, nor that it contradicts my training as a physicist, which taught me to never stop trying to understand the world. As a concept itself, queerness is complex, at times confusing; its meaning and manifestations are always evolving and shifting. It can provide comfort and stability, but it does not render impossible or unimportant the moments when I am questioned or perturbed, when I have to rethink and rearticulate my identity again and again.

Karen Barad argues that nature can't be treated as something amenable to being chunked up into separate, passive parts and experimented upon. For them, an electron materializes into reality through a communal effort of the experimenter, the tools of the experiment, and the electron itself. Barad calls this an "intra-action" and takes it to be constitutive, in some way an act of creation of whatever reality it is set up to create. They use the word "intra-action" instead of "inter-action" to emphasize that everything comes into being through its connections with something else, through an entanglement with the rest of the world that it is a part of.[30]

This is a philosophical view, and it is not mainstream among academic physicists. But I find it compelling because it reminds me that although I have found a word to describe myself, I still need to keep thinking about how I move through the world and how I interact with others. As the writer Torrey Peters says, gender, among everything else that it is, "is also a negotiation with other people."[31] Being queer is a space that I am called on to explore and spread across—it is not a well-defined point on which I can stand definitively, in perfectly certainty, acting as though I'd never learned that classical physics was misguided in making that same assertion.

Building on the queer theorist José Esteban Muñoz, who wrote of queerness as "the warm illumination of a horizon imbued with potentiality," Prescod-Weinstein asserts that "queerness is conceptually always at the edge of knowledge."[32] Had the hunger for that knowledge not been ingrained in my queer self, I would have probably never become a physicist.

CHAPTER 11

ELECTROMAGNETICALLY INDUCED TRANSPARENCY

CLASS ASSIGNMENT 2

Our goal today is to understand, and be able to explain, what electromagnetically induced transparency (EIT) is, and how it may be explored in laboratory experiments.

1. Let's break down the words in that phrase first. List some transparent objects. What would you do to prove to someone that they are transparent? (If you get stuck, look up some synonyms for "transparent.")
2. Informed by your list, write a rough definition for what it means for an object to be transparent. It may be helpful to mention light.
3. If something is "electromagnetically induced," that means that some combination of electric and magnetic forces made that something happen. Write down an example of when you experienced electric or magnetic forces.
4. We can now translate the whole phrase: "Electromagnetically induced transparency" is when you use electric and magnetic forces to turn an object transparent.

Before I sat in classrooms as a student, I visited them as the teacher's kid. My mother is a lifelong teacher, and watching her do her job was my first lesson in how teachers maintain power over what counts as knowledge and truth. I would watch my mom watch her students' presentations and see how she would occasionally nod or interject, signaling to the class what facts they should absorb. The idea of being someone who decrees what is correct enough to be committed to memory, and then folded into a framework for how the world works, appealed to me—I only gave up on being a teacher after I realized that if I became a scientist, I could uncover and create knowledge instead. But after I moved to the United States to become a physicist, for more than a decade, all of my jobs ended up being teaching positions.

In college, being a teaching assistant was the only job that I could legally hold while on my student visa, so I spent years grading homework and holding office hours in coffee shops. I was no profound truth-teller, but I relished being a person with expertise and authority. In the summers, the research fellowships that my American peers took were not available to me as a foreign national, so I kept teaching. This pattern continued in graduate school, where for most of those six and a half years I was always someone's teaching assistant.

By the time I earned my PhD, I had facilitated several different types of problem-solving sessions and labs, written dozens of quizzes and homework problems, graded hundreds of each, and interacted with countless students representing a fairly broad diversity of backgrounds. I strove to be rigorous and fair but still likable, sort of a personable hard-ass, like my own favorite professors. Sometimes my teaching reviews described me as arrogant, but I mostly got along with my students, and the older I got, the more they seemed to respect my commitment to physics—a discipline they often saw as both profound and difficult. Because teaching had become routine for me, I was also rather confident that I could handle almost any college classroom, and that confidence felt good.

But I was still only formally trained as a researcher, and I thought of myself as a researcher to such an extent that I could not imagine ever being a teacher only. Certainly, I never had imagined choosing to do so while sheltering in place because of a nascent global pandemic.

In the spring of 2020, however, I was a newly minted doctor of physics, I had no leads on academic research jobs, and I really needed to work. It was a stressful and uncertain time, so I felt the need to play it safe, which led me back to the one job that had always been there for me. In September, I started as a faculty member at a competitive New York City high school, and I was terrified of what was ahead of me.

CLASS ASSIGNMENT 3

We say that household objects like glass and plastic wrap are "transparent" when we can see through them. (In everyday language, "see-through" and "transparent" are synonyms.)

Recall our discussion of eyesight: when we see an object it is because light hits it, then enters our eyes. Here are some more specific scenarios:

1. If particles of light get absorbed by the object they hit, so that they get stuck inside it instead of traveling to our eyes, that object appears dark.
2. If all the particles of light bounce off the object, a lot of light enters our eyes at once, and that object appears bright.
3. If some colors of light get absorbed and some reflected, we see the object as having the same color as the reflected light.

What do you think happens to particles of light when the object is transparent?

When a particle of light hits a transparent object, I think it ______.

Trying to do anything in 2020 was confusing and unpredictable, but schools were especially swept up in the chaos of the pandemic. Class attendance rules kept changing. School buildings closed, then reopened, then closed again. Everyone was constantly scrambling to adjust to the next "new normal." I started the school year over Zoom, speaking into a computer screen crowded with scared faces, while living in my in-laws' basement. I gave my final lecture of the year in a socially distanced classroom, through two masks, flanked by windows that I always kept open. The smallest sign of illness—a sneeze or a sniffle—scared everyone, as did any unexpected schoolwide announcements from the principal. And there were still always so many lessons to get through.

Though I can, in hindsight, chalk up a lot of my teaching anxieties to COVID-19, in the moment I was also deeply scared by not just the amount of authority I'd been given, but also the amount of responsibility. Right from the very first lesson—the one titled "What Is Physics?"—I felt underprepared for it all. I had spent all my adult life living, talking, dreaming, doing, and teaching physics, but none of that made me feel worthy of being a fifteen-year-old's first physicist—their first portal into this storied science. Even more worrying, I had never before considered how to convince students that physics matters, because in all of my past classrooms, I could safely assume that my students were majoring either in physics or a related science and thus already thought it valuable. At this high school, my class was mandatory for all ninth graders, but the school's prestigious reputation was not rooted in the sciences. On the contrary, it was the sort of place where students went to read big books or take classes on law or advanced literature at a young age. A few weeks into the job, I was even told a very hyperbolic anecdote about a former physics teacher at the school who had effectively stopped doing their job because they felt the students did not care enough about the subject.

Faced with the prospect of my class being an afterthought or an annoyance, I could have dug my heels in, stood on my authority as the adult in the room, and just asserted that physics matters, and matters a lot. But nothing about that felt right. It felt unjust, and in direct opposition to the spirit of doing science, much of which is about repeatedly asking "why." Such an absolutist and authoritarian approach would have also been especially fraught in a year when the pandemic had made trust in science both endangered and extremely important.

The textbooks, including those recommended by some of my new colleagues, were of no help. These books offered a story of physics that was more like a fairy tale than anything I had experienced. Here, the steps of the scientific method were always performed neatly and in order, and they always worked. When researchers set out to interpret experimental findings, they ended up with absolute truths, which they then communicated clearly and precisely. When contradictory evidence arose, they always changed their views in the end. Everyone only ever argued in good faith, everyone's ideas were considered regardless of who they were, and scientific objectivity always triumphed over political influences. There were no structural or societal inequalities in the world of these textbooks. Maybe you got the story of Galileo Galilei fighting the Catholic Church in the seventeenth century, a neat, nearly mythical story about a man martyred for the sake of rationality and uncomplicated progress. In textbooks, physicists seemed to be perfectly objective and to care only about upholding a level of rigor that, the textbooks said, put all other sciences to shame.

The world of physics that I had experienced was, by contrast, as messy and fallible and joyful as the people who practiced it. Even the researchers I most admired constantly worried about securing funding, reaching promotion goals, and maintaining prestige. Though many had an abundance of creativity and passion, the university system in which they worked made it practically impossible to center their work solely on those passions, or to do it at the sometimes painstakingly slow natural pace of thought and discovery.

Universities are run like businesses, and I'd seen how that mindset also made it more tempting for physicists to make ethically dubious decisions, such as accepting military funding, or looking the other way when their colleagues and students worked hours long enough to damage their health. I'd certainly seen physicists encourage and even praise this kind of self-destructive behavior, presumably in the service of producing more scientific breakthroughs more quickly. In quantum physics research, there is an increasing trend of spinning start-ups out of university labs, directly connecting academic labor to a work culture more reminiscent of Silicon Valley.

Nafis Hasan writes about a similar tendency in cancer and other medical research: "The neoliberalization of scientific research has turned every single biomedical scientist into an entrepreneur, every single lab into a small business opportunity."[1]

The most prestigious physics journals still very rarely publish papers on methods that did not work, or that failed to replicate others' results.

This skews researchers' perceptions of what types of scientific work are worthwhile, and it tempts them to oversell their own discoveries. The style in which papers are written has become so specialized that even when claims of discovery are not inflated, they are inaccessible to everyone but a handful of experts. Being an expert also often means having enough social capital to refuse requests to share data—and to critique the work of others based on interpersonal conflict and bias instead of merit.

As a science journalist today, I hear researchers complain about this constantly. They go off the record to explain why they may not want to comment on a colleague's work, or why they have no time, or focus, to read others' papers. These conversations sadden me, but they have long been familiar. For me, it was always curiosity that propelled me through academia, but it took me longer to learn that to be a successful physicist you also needed a thick skin, a strong support system, and a good amount of luck. If you had all of that, and your technical skills were strong, you still had to navigate the culture of physics, which has long been shaped by maleness, whiteness, and colonialism. Many talented researchers have found this to be too tall a hill to climb, while others have made the climb but found themselves changed and jaded at the top. The textbooks I was meant to give my students conveniently ignored this.

This is not to say that physics does not work—all the devices that we have around us, and the fact that we know why the sun keeps setting and rising, confirm that it does. Nor is it to say that physicists are not driven by a fascination with the world. But once you lift the veil of the stories we tell about physics, you do find a very messy, complicated reality. Sadly, many stories about physics also tend to downplay how joyous physics can be, how much love many physicists have for the oddness of nature, how much creativity and art can exist in their work, and how much they depend on each other. Centering only results, productivity, or technical difficulties that had to be overcome erases some of physicists' basic humanity. Textbooks almost always give us individual success stories—lone geniuses and an occasional underdog. Consequently, they also erase so much of the vibrancy of doing physics as a community.

I was at a loss on how to convey these complications and tensions to my students. Could any of it be taught at all?

CLASS ASSIGNMENT 4

We arrived at the idea that whether an object is transparent depends on if a particle of light can pass through it. Each particle carries some energy. What do you think happens to that energy when the particle doesn't make it out of the object?

Useful facts:

1. Energy is conserved (look at your past notes if you don't remember what this means).
2. If you leave a dark-colored sweater in the sun, it gets warm.

In the *Encyclopedia of Science Education*, the education researcher Steve Alsop writes, "As science educators, our pedagogies are at their heart an invitation to invite others into our worlds and experience a subject that has occupied our minds and emotions for such a long time. This invitation carries with it an open prospect of encountering the wonderment, delight, hopes, challenges, and possibilities of seeing the world and ourselves in different ways."[2]

Before I started teaching ninth graders, the idea of inviting students into the world of physics that I had built within myself seemed misguided. I worried that if I ever got too honest, if my views became too transparent, I would undermine my authority and detract from the value of the content that I was teaching. I felt like I needed to be opaque, a black box that takes questions and homework as inputs and then spews out lectures and grades.

But in 2020, times were so strange that novelty and recklessness were almost expected. My employer gave me lots of freedom and minimal mentorship, and I was coming to terms with my research career being over. Nothing seemed to be going according to plan, for me or the world, and the least I could do, I started to tell myself, was show my students what physics can be like when you do it, and love it, for real.

So, during the first week of classes, even before I talked about physics, I talked about myself.

"I want you to understand who I am and what I bring into the classroom," I said while showing slides with titles like "Who I Am" and "How I Got Here." In the social sciences, this type of self-disclosure is called "positionality," and at the time I had only known labor and equity organizers to practice it, in anticipation of difficult conversations that would require lots of trust. Now, I thought that a fairly heterogeneous group composed of students and me that was trying to study the fundamental nature of physical reality could benefit from the same.

On the first page of the syllabus, I posed a difficult question. "Why should you care about this course?" asked the thick purple letters. Underneath that question, I made my case to the class, arguing:

> We will strive to practice thinking like physicists: examining evidence and data, approaching problems in a systematic and analytic way, testing hypotheses, engaging in discussions with others and welcoming challenges, noticing patterns and organizing our thoughts into internally consistent theories. This way of thinking will bring us closer to understanding how working scientists do what they do, and why they think they know what they know. It will also be good practice for all other areas in which you will encounter complex and complicated problems.

More earnestly, the last page read:

> Physics can be hard and at some point, everyone that studies physics struggles. Often, the science we see reported on in the media or written about in books focuses on successful experiments and breakthroughs in theories. In reality, behind all of those successes is a whole lot of failure and frustration that we don't necessarily get to see. It is normal to sometimes do badly or be confused or fall behind. No one is born a physicist and there is no physics gene that makes some people better at physics than others.

My "What Is Physics?" lesson included lots of definitions and examples, as well as an activity where students explored the websites of physics departments at big research universities, and I reserved time for a discussion of who physicists are. In 2021, for instance, 76 percent of all recipients of a physics bachelor's degree were men, and 58 percent of all recipients were white.[3] I presented this not to be sanctimonious or assign blame, but to offer context about who exactly is using all the laws and equations from our curriculum to deduce and tell the story of our shared universe.

During Black History Month, I showed interviews with Black physicists who discussed their research and their experience of being underrepresented in their field. During Pride Month, I did the same with several queer physicists. It was important to me that we discuss these researchers' challenges and successes alike, and I sat in the discomfort of claiming that anyone can strive to be a physicist while also acknowledging that, for some people, the barriers to entry are very high. I had lived a version of this, and some of my students, especially the seniors in my more advanced classes, asked about it directly. I spoke as honestly as I could, resisting the urge to hide the hard stuff just to make the good stuff seem more worthwhile.

It was a year of racially charged reckonings, and the moral panic over queer teachers—which has since resulted in lost jobs and banned books—was already in full swing, so some of these topics also felt dangerous. I worried that letting students see through me would put me in the crosshairs of school administrators or conservative parents. Again, I resolved to be honest, and was heartened to find my students largely open-minded and curious. After all, they were already part of an academic system that taught them a lot but also othered many of them in the process.

Whether by design or the result of oversight, many of my students were aware that their identities affected the conditions and the perception of their work. Interacting with students and teachers who were very different from them had also shown them that their identities and origins had shaped their sense of the world. Our conversations encouraged me to not shy away from complex ideas, scientific or societal, and to keep trying to engage with these personal experiences.

CLASS ASSIGNMENT 5

Let's take stock of what we have so far.

1. Objects are transparent when particles of light can pass through them.
2. When a particle of light hits an object, its energy cannot be destroyed, according to the law of conservation of energy.
3. Objects that absorb light appear dark, because light never makes it into our eyes, and they get warmed because they take on the energy that the light was carrying.
4. Transparent objects, then, let the particle pass through because they do not absorb much or any of its energy.

This leads to a new definition: an object that does not absorb too much energy from the light is likely to appear see-through.

Here's a follow-up question to help you assess your understanding:

Q: Suppose you have two lasers, a red one and a green one, and each color of light carries a slightly different amount of energy, E_R and E_G. Your labmate gives you a crystal that you've never seen before and explains that it has one special property: the only amount of energy it can fully absorb is E_R. When you hold it up to the light, it mostly looks blue. What might happen if you point your green laser onto this crystal?

a. The green laser light will pass through the crystal, so the crystal will be transparent to green light.
b. The crystal won't absorb the green laser light, but it might reflect it.

The idea of teaching authentic science also greatly influenced my work. As Barbara Crawford describes it, "Authentic refers to using data and logic to create an explanation for something not known or understood and using skepticism about the best explanations or applications." This approach stands in contrast to more contrived modes of problem-solving where students do very little discovery and invention, and instead just follow the predetermined steps of a calculation or a laboratory experiment.[4] In practical terms, teaching authentic science means designing assignments and exercises that resemble the work of practicing scientists—and empowering students to take charge of how they deduce and discover knowledge.

I loved this idea, even though I sometimes ran into trouble when trying to implement it. In one notable example, I wanted students to discover the formula for the velocity of an object—in this case, a battery-powered toy car. Divided into groups, equipped with rulers, stopwatches, and toy cars, the students tried to work out how they could use the quantities they were able to measure—time and distance—to calculate velocity. Many did figure out that the trick was dividing the distance that the car traveled by the time it took to travel it, but one group ended up with a faulty toy that started to smoke. The moment when a student approached me to very politely say, "Dr. Callaghan, I think my car is on fire," still features in my anxiety dreams. But I managed to put that fire out, as well as many other less literal ones, always reminding myself that the physics I had lived as a researcher had been full of awkward moments too. In some sense, very little could be more authentic than struggling with lab equipment and failing.

But I wanted to apply the idea of authenticity to more than what physics teaches us—I also wanted to use it to shape how my students encountered physics itself. I had often been told that it was important to learn how to "think as a physicist," but I also wanted to remember the best parts of how I'd learned to "be like a physicist."

Almost all the physicists I knew worked as part of a research team, so in my introductory courses, each student was assigned a "designated collaborator." The two would check in once a week and ask each other questions like, "What stood out to you the most in class this week?" and "What was most confusing for you?" Sometimes they helped each other gain clarity, sometimes they just commiserated. Both seemed beneficial to me and, playing the role of a "principal investigator," I returned their notes with a layer of my own, then chatted with them about it in class. "Science is a team sport," I repeated every time we huddled to take stock of how our shared learning project was going. "You get to decide how you show up and what you contribute."

I learned a lot about best teaching practices from books and workshops, and from observing colleagues who had gotten their degrees in education rather than physics. But once I started seeing parallels between my experiences as a researcher and what I was finding to be effective in the classroom, it was hard to ignore them. Not running away from the fact that I had been a researcher, and an imperfect one at that, once again proved more valuable than hiding behind some opaque mask of an all-knowing teacher.

CLASS ASSIGNMENT 6

> What about the first two words of "electromagnetically induced transparency"?

Earlier, we said "electromagnetically induced transparency" is when you use electric and magnetic forces to turn an object transparent, but now we know that whether an object is transparent depends on what energies it can absorb. Incorporate that insight into the statement below:

> "Electromagnetically induced transparency is when you use electric and magnetic forces to change ______ that an object can ______."

In my second year as a teacher, I started to encourage students to embrace failure as the most common first step toward success. When I asked my students to compile advice for the group that would succeed them, someone very frankly wrote, "It's okay to be bad at physics because KPC gives you second chances." In fact, I had redesigned the class so that it became nearly impossible to get a failing grade and second chances were mandatory. Students were required to keep retaking quizzes and exams until they could demonstrate enough understanding for at least a C, instead of settling for a D or letting themselves fail so they'd never have to think about the topic again. Joe Feldman, in *Grading for Equity: What It Is, Why It Matters, and How It Can Transform Schools and Classrooms*, argues that this approach makes grading fairer because it focuses on the student eventually acquiring skills rather than punishing them for not doing so immediately.[5] It also matches the experience of doing research in the real world. The real world is often used as a boogeyman against students who ask for second chances, but in the field of scientific research, most days are spent failing, and most days serve as the second, third, or whatever chance you need to achieve your goal.

I talked about this often, reassuring students that it had taken me three years of work to publish my first research paper in graduate school, and that in my first quantum field theory class, I would still get wrong answers on homework problems even after having worked on them for eight or twelve hours. "I know it feels terrible, but sometimes you can only learn by doing hard things," I would say. In one way or another, I repeatedly told them that it's just good to want to know things, and to keep trying to learn them. This belief had kept me going through my toughest times in graduate school, and now it kept me coming to work and entering the classroom with as much energy as I could muster—even when the rest of the world felt extremely grim.

Gradually, except for a few times a year when I had to submit them to the school's administration, I gave up on letter grades and instead started giving out report cards that read like a checklist of skills. "You are absolutely ready to break down any problem according to the steps of the scientific method," and "You really understand how to determine an object's velocity," I would write. I never worried about students trying to "game the system," because their grades were never a reward—just another way to take stock of their progress. I handed out rubrics with every single exam so my assessment methods would be fully transparent. Often, I asked students to research some small portion of the lesson and teach it to each other, so they could get a sense of what it was like to not just find information, but also communicate it and identify important takeaways, like I constantly had to do.

Because I have always been chronically online and recognized in my students a shared instinct to reach for our phones, we discussed googling scientific facts, and I showed them the websites I used whenever I needed to quickly refresh my memory. My twelfth-grade students thought that I was joking when I said that "I've only memorized like three equations in my whole physics career," but I saw them beam with pride when they learned to derive a formula instead of rummaging for it through their memory.

Still, I asked them to write "cheat sheets" that listed all the useful formulas for use during their exams, always telling them how much reference material I'd needed for every single piece of research I'd done. Some of the other science teachers in the school had stricter policies on formula sheets, requiring them to be shorter, with no annotations allowed. I once would have worried about those colleagues seeing my approach as sloppy, or less than, but now I believed that my students and I were simply practicing science on our own terms, authentically. I was a better teacher when I was not trying to perform the persona of some sort of textbook lone genius.

My upper-level classes featured as many problem sessions—in which students spent the whole period working on a set of problems in small groups—as they did lectures. I'd walk around and ask the kinds of questions that my research adviser used to ask me: "Can you tell me the purpose of this calculation step? Why did you choose this particular equation? Did you start with a piece of information or an assumption?" I let students falter and stumble over their words for as long as they needed to, promising them that talking about their work as their own science practice, and not just mine, would get easier over time. I constantly asked students to present on something, or write something on the board, or yell it out at me when we were on Zoom. After years of being something of a self-imposed curmudgeon, now I was cheerleading and bouncing with energy, even when this meant ending the day with tears of exhaustion catching in my throat.

I put a lot of effort into being transparent about the kind of scientist I was, and how I became one. But I also wanted to recognize that, as the education researcher Lucy Avraamidou puts it, there is a "landscape of the infinite ways of becoming a science person."[6] Discussing who physicists are and are not, practicing physics the way physicists do, being honest about the strengths and shortcomings of the field, projecting the personal rather than some idealized objectivity—all this created an atmosphere in which, I hoped, everyone could engage with physics in the way that made sense for them, and become an enthusiastic stakeholder in telling the story of our physical reality. So many of my students entered the classroom convinced that they were simply not "a science person." I wanted them to leave it feeling like they had, inadvertently, at least a little, become one.

CLASS ASSIGNMENT 7

> Do you think we can actually make some object transparent by controlling if its atoms will take in energy or reject it?

Look this up on the Internet, focusing on information about experiments. It's okay if you come across a lot of words that you don't know. Take notes on what you find, and we will unpack them together.

Hint: If you find the words "quantum" and "ultracold," you're on the right track.

I could tell you that I was a successful teacher by pointing to how many of my students earned high scores on their exams, or to the students who became science majors in college despite not having attended a STEM-centric high school, or to the parents who told me they were shocked by their kid's enthusiasm for our class. I could tell you that when I decided to leave teaching, my students cried, and the school administrators negotiated to keep me in the classroom as long as possible. I could tell you that a former student once messaged me to say that they were writing a paper in their college gender studies course about the classroom experiences we'd shared, and that others still want me to write them recommendations for leadership positions and summer fellowships. These are all accomplishments, and they feel validating.

But what I really want to tell you is that after I decided to make myself transparent to my students, to let them see through the layers of authority and expertise that I had accumulated during my years as an academic, it became easier for me to recognize their shine. I would catch glimmers of their curiosity, and their radiance in the joy of discovery. Their bright light felt good, like it illuminated a path toward a future that I would have otherwise been scared to imagine. And I could, with this light, also look into myself and see that I had changed too. So many of my fears about lacking, or losing, authority had fallen away, and my love of science—and the world whose story it tells—had grown. My anxiety about explaining what physics is, and why it matters, was finally extinguished now that I understood that the key to it all was to just be human and share physics instead.

CLASS ASSIGNMENT 8

Let's summarize our exploration. You can write in the style of a textbook or as if you were describing something you did in the lab, you can make a list, or you can draw a series of images. Once you are finished, flip the page and read my summary. Underline any parts where our summaries differ. When everyone completes this step, we will read around the room and discuss. If your peers still seem to be working once you have finished, please raise your hand.

Dr. Callaghan's summary:

If I had to work in a lab that turns objects transparent, I would choose a lab where other researchers were really good at dealing with quantum objects, like extremely cold atoms. This is because those objects can be controlled very precisely, not just in terms of how they move, but also in terms of what happens inside of them. Specifically, you can use electromagnetic forces to change what happens inside of an ultracold atom when a particle of light that carries some specific energy hits it. This is what I would do: I would determine the energy of the light I wanted to use, then ask my labmates to expose their collection of atoms to electric and magnetic forces that would tweak the atom's insides just so that energy does not get absorbed or reflected. Then, I would shine my light onto the atoms, and with their internal structure just right, I would see them become transparent.

CHAPTER 12

QUANTUM SPIN

To do quantum physics, you have to learn a lot of special words. There are new words like "fermions" and "bosons," there are unexpectedly whimsical phrases like "ghost fields" and "magic states," and there are words that tickle your intuition without revealing their full meaning, like "superfluid." But one of the most difficult-to-understand words in quantum physics sounds tantalizingly self-explanatory: "spin."

All quantum particles have spin, which is either represented as an integer, like 0 or 1, or a fraction, like 1/2. But what physical property these numbers represent is a challenging question. This is because researchers have agreed, for almost as long as they have known that spin exists, that none of these particles are literally spinning. In fact, our best theories all suggest that such spinning would be impossible.

This is confusing. One look at online message boards finds physics students and enthusiasts alike seeking clarification. In a thread on what spin is, one user on the Ask Physics subreddit says it directly: "I swear this question comes up monthly."[1] Another offers an explanation that is not kind but also not incorrect: "The orthodox answer is generally that [the particle] is not spinning, it just has this property that just happens to have units of angular momentum and behaves like angular momentum, and is angular momentum, and shut up and calculate." A third reminds me of how I thought about spin as a physicist: "Don't try to think of it as coming from an actual rotation, that will end in tears. Just try to understand it mathematically."

I used to find some comfort in this approach of forgoing words and embracing mathematics. I could write down a matrix, a two-by-two grid of symbols, and dismiss all discussion as something that was merely steering me away from precision. My first quantum mechanics textbook, the mathematically advanced one, by Shankar, with the iconic red cover, also encouraged this mindset:

"This angular momentum is called spin, for it was imagined in the early days that if the electron has angular momentum without moving through space, then it must be spinning like a top. We adopt this nomenclature, but not the mechanical model that goes with it, for a consistent mechanical model doesn't exist. Fortunately, one can describe spin and its dynamics without appealing to any model."[2]

A fully mathematical description follows. But even for Shankar, this is not enough. He very quickly brings you back to the tangible physical world by connecting spin to a famous 1922 experiment with silver atoms and magnets. This experiment, named after the physicists Otto Stern and Walther Gerlach, is hailed as a paradigmatic example of how spin manifests in the real world. (More on that later.) So, while Shankar cannot offer you an image or a diagram of spin, he does what most quantum theory texts do and provides this experiment as something you can visualize instead. Mathematics alone does not actually cut it.

This is not surprising: physics is an empirical discipline, so any abstract concept must be tied to an experiment in order to be fully meaningful. But with spin, the experimental evidence feels more essential, because the thing it is demonstrating seems to lack any proper representation beyond mathematics. Moreover, though we can't really find the right words for what spin is, it is nonetheless key to understanding why our world is the way it is. For one, the spin of particles like electrons explains how magnets work. More crucially, the existence of electrons with the kind of spin that we represent with the number 1/2 explains why every solid object you have ever touched, bumped into, or sat on could be solid to begin with.

So, spin matters an awful lot, and physicists are pretty certain that the way they fold it into their quantum theory equations is correct—it has matched experimental observation for about a century. Yet, every time someone says "the spin of the electron," the image that phrase suggests is absolutely incorrect.

When physicists speak to each other, there is a shared understanding of this error, so they rarely qualify their use of the word "spin." The error has become normalized because it is assumed that everyone knows about it and knows to catch themselves from propagating it into their work.

I did this too when I was a physicist. It didn't worry me that I was lying about the nature of electrons, because everyone was doing it too, and if the lie got to be too much, we always still had math. The vocabulary of physics largely predates quantum theory, so it can genuinely feel like a losing game to be describing quantum phenomena in fundamentally non-quantum language. But then I became a science writer, and that losing game became the only game in town for me.

Though I only became a professional writer in my thirties, I'd always had some ongoing writing project, whether it was in booklets handbound by my mom when I was young or, later, on the Internet. When I started graduate school in physics, I worried that I'd have to leave words behind and fully devote my time to the language of mathematics. Then, I learned that to be a successful physicist, you actually have to think about writing words constantly.

Although my classes assigned math-heavy homework, and although I did have to pass a long qualifying exam full of calculations, my output as a researcher was almost always evaluated only after I'd converted it to writing. This reflects the experience of more established physicists, as their most quantifiable output also takes the form of papers. The point of doing experiments or solving equations is to discover something new about the physical world, but the way you announce that discovery—the way you make it possible for your colleagues to double-check you and show your funders that you put their money to good use—is primarily by writing papers, then publishing them in peer-reviewed journals. Within my research group, the unspoken rule was that students should coauthor at least four papers before graduating. My research adviser also had to write grant proposals, which, when successful, kept the lights on for all of us.

So, when I was a graduate student, choosing the right words to describe and explain my work was never far from my mind. Studies show that this is a common experience for most academic physicists.[3] The average number of physics papers published per year has been around twenty-one thousand for most of the past century, and most physicists coauthor at least one paper a year. And these are conservative estimates, not necessarily inclusive of physicists who work and publish in more interdisciplinary ways.

I was listed as coauthor on two peer-reviewed papers in college, but it was actually just one graduate student from my research group who wrote nearly all of both. I was impressed by how he transformed our endless, loopy meetings, full of theorizing and half-baked ideas, into something that sounded so precise and authoritative. It was much harder to follow the discussion on the page, in the language of an academic journal, than when we talked face-to-face, but I also couldn't help thinking that my work sounded so much more like a deep truth about the physical world when it was published.

A few years later, I was that graduate student who was writing nearly everything. I had to not only observe and digest the academic journal style, but also learn how to use it. The great tool that turned unruly thoughts into precise truths was now in my hands—and I struggled with it. I found myself running out of space, running out of words, paralyzed by the idea that I might end up sounding too subjective, too idiosyncratic, too casual. I started to worry that the style that most journals preferred could hamper my ability to communicate if I got it wrong—and even if I got it right.

For one, most of the prestigious journals favor shorter papers, which forces you to summarize years of work in as little as four pages and push most of the details into dry and technical "supplemental materials." Often, you also sacrifice any sort of personal flair in writing—or what I now, as a writer, know to call "voice." Many researchers have undeniably distinct voices in person but, on the page, they end up relying on stock phrases and jargon. I was trained by a physicist with a bright personality and a large capacity, and affinity, for beautiful prose, but only tiny glimmers of that showed in the papers that we wrote together.

In the competition between objective, mechanical precision and descriptive, subjective language, many journal editors will heavily tip the scale toward the former. Certainly, impartiality is crucial for the success of physics as a mode of meaning-making, but there are also less idealistic reasons for this writing style. For instance, the desire to maintain a journal's prestige by making it clear that its content is so advanced that only a narrow set of experts can comprehend it.

This is to the detriment of both physics and physicists, as the physicist David Mermin wrote over thirty years ago. Mermin, who led a successful career as both a researcher and a science communicator, argued that "the insistence on bland impersonality and the widespread indifference to anything like the display of a unique human author in scientific exposition have not only transformed the reading of most scientific papers into an act of tedious drudgery, but have also deprived scientists of some powerful tools for enhancing their clarity in communicating matters of great complexity."[4]

Papers from past eras of physics, like those written by the founders of quantum mechanics, are a lot more fun to read. In this way, they reflect their time—their authors existed in a different publishing system, as well as a different system of work and employment. Today, more and more universities are run like businesses, and academic publishing itself is certainly a full-fledged business. Scientists who were at some point driven by pure curiosity find themselves trying to fulfill more and more metrics, manage their own students as if that were a business too, and optimize their productivity, regardless of whether the natural pace of discovery can actually be rushed.

In some sense, physicists are being pushed toward becoming workers instead of visionaries or dreamers. That puts beautiful writing out of their minds even before they confront the possibility of rejection by a journal editor. They start to see writing as another chore, and words as another transactional part of their career, rather than an enticing entry point for understanding the intricacies of physical reality. It can be hard to catch yourself losing sight of how important it is for words to feel more personal, to feel like you've taken them and their encoded ideas with you into the shower and put them under your pillow when you go to sleep.

After I left academic research to become a teacher, I constantly had to negotiate with myself over what words I would use in the classroom. My instinct was to teach students jargon so they could read papers and absorb precise statements about the physical world. Repeatedly, I had to force myself to stop and ask whether there was a way for them to understand the truths of that world in a way that was more accessible, less robotic and rote. Now that I work as a journalist, this sometimes feels like my whole job.

I write hundreds of popular science articles a year, and most of them start with me reading a paper in a journal. The beats and conventions of the academic journal style still feel familiar—I did learn it once, after all, and I did believe that it would help me write accurately and correctly. When I search for correctness now, however, I most often find myself trying to get on the phone with the researchers and asking them to give me words that actually feel lived in—and human.

In 1922, Walther Gerlach and Otto Stern performed the experiment that made visible the consequences of spin. They also had no idea that spin existed. In fact, five more years would pass before anyone would use the word "spin" to explain what the duo had seen in the lab.

Really, it all started because Stern was skeptical of quantum theory and, specifically, of Niels Bohr's atomic model. He told a colleague, "If this nonsense of Bohr's should in the end prove to be right, we will quit physics!"[5] Since he was a physicist, the only way he knew how to ascertain whether Bohr was wrong was through an experiment. The story goes that he thought it up while lying down one morning, as "it was too cold to get out of bed."[6]

In Bohr's model of the atom, electrons move around the nucleus in orbits, like planets around the sun. To explain why this structure can produce a stable atom, or what exactly keeps the electrons from falling out of their orbits and crashing into the nucleus, Bohr and his colleagues introduced the idea of "quantization."

A physical property that is quantized can only take values from a discrete set of numbers. Imagine filling a glass with water. Ordinarily, you can stop pouring water into the glass whenever you want, at any volume. But if the volume of the water were quantized, you could fill it up, for example, to one cup, or two cups, or three cups, but never anything in between. If you tried to fill the glass with one and a half cups of water, the extraneous half would be forced out and splash onto you. This is what Bohr and some of his contemporaries suggested was happening within the atom—that there were discrete amounts of rotation that an electron could have, and each corresponded to a stable orbit around the nucleus. If an electron had more energy than one of those special amounts, it would release it as a splash of light and "fall" into a lower orbit.

For the most part, the data from spectroscopy experiments, from which these special energies were inferred based on light that atoms emitted after being warmed, supported this model. Except for one feature. It seemed that, within some atoms, the electron with the lowest discrete energy suffered an extra restriction—the direction of its rotation could only take two values. This became known as "space quantization." Stern wanted to bring this esoteric phrase closer to the real world by connecting it to the phenomenon of magnetism.

He knew that if you made a rotating charged particle fly near a magnet, the magnet would exert a force on the particle and bend its trajectory. If Bohr's lowest-energy electrons really only had two options for how to rotate, then this magnetic force would have only two effects as well. A magnet would split a beam of atoms into two smaller beams, each pointing in different directions, because atoms with opposite values of space quantization would have their trajectories bent in opposite directions. This was the gist of Stern's experiment.

Working with Walther Gerlach, he constructed a setup where a beam of silver atoms hit a screen after traveling between some magnets. The researchers hoped that they could then look at the screen and literally see whether quantum theory was accurate. If the theory was incorrect, they would see all the silver hit the screen close to the same point, forming a sort of diffuse dark blob. In this case, Stern would have been proven correct in his skepticism. If the theory was correct, however, the blob would be split into two: the atoms would make one set of marks for each of the two values of space quantization. In February of 1922, Bohr received a postcard from Gerlach. It showed a photograph of the screen clearly displaying that the atom beam had split in two and made two different marks.

The Stern-Gerlach experiment seemed to have proven that space quantization was real. But that did not mean that physicists had precise words for what "space quantization" meant. The physicist Max Born, who made important contributions to quantum theory himself, wrote in his autobiography in 1978 that he'd "thought always that [space] quantization was a kind of symbolic expression for something which you don't understand."[7] Another impactful physicist of the time, Peter Debye, told Gerlach while he was working on the experiment, "Surely you don't believe that the [spatial] orientation of atoms is something physically real; that is [only] a timetable for the electrons."[8]

Would Born and Debye have felt like the property that Stern and Gerlach were after was more concrete and intuitively physical had they known to call it spin? When today's physicists talk about spin—beyond any discussion of the fallacy of the word itself—they sound very similar to how Debye and Born reacted to "space quantization." This is a pattern that repeats throughout the history of quantum mechanics. Researchers can present the precision, logic, and internal consistency of their mathematical work, and they can present their tangible, reproducible, intentionally constructed experiments. Yet, they cannot always produce words and images that match either with perfect accuracy.

In 1958, Werner Heisenberg stated this directly in his writings on quantum theory and philosophy. While discussing the new phenomena that physicists could now explore in experiments because of technological advances, he wondered about the best way to describe those phenomena. He acknowledged that "the first language that emerges from the process of scientific clarification is in theoretical physics, usually a mathematical language, the mathematical scheme, which allows one to predict the results of experiments," and that "the physicist may be satisfied when he has the mathematical scheme and knows how to use it for the interpretation of the experiments." But this is not enough, Heisenberg notes: the physicist must also "speak about his results also to non-physicists who will not be satisfied unless some explanation is given in plain language, understandable to anybody. Even for the physicist the description in plain language will be a criterion of the degree of understanding that has been reached."[9]

In the case of "spin," it is unclear whether physicists have ever really met that criterion.

Learning to write as a physicist taught me that words really matter, and not just for appeasing journal editors. They have definitions, which leave little wiggle room for imprecision, and they correspond to mathematical equations, which are always there to push that level of precision further. In my PhD thesis, every word contributed to a picture of the world that I was trying to assemble based on my research findings, and when I presented some of those findings at conferences, other physicists were quick to scrutinize them.

"You can't just say things," my quantum field theory (QFT) professor once admonished a student during a final project presentation. It's such a simple, almost banal consideration, yet a crucial one. How much work do you have to do before you can use a word, before you earn the privilege of employing its precision, or even earn the authority to underscore its limitations? Now that I am writing for people who are not physicists, whenever I introduce a new word, I also have to ask myself how it serves the reader. These words don't just matter for me and a handful of other researchers who care about my niche topic—they're also for thousands of lay readers for whom my writing may be an entry point into physics overall.

For people who may never crack open an academic journal and do not worry about physics' deeply mathematical bones, words matter in a whole other set of ways. They are still vehicles for meaning, but they often also have to be handholds through rocky reasoning and, at their best, they are a soft or exuberant invitation into something new. Too much jargon can put up a tall wall of effort between the reader and the story—a wall that, ironically, obscures the very precise meanings that those jargon words were invented to convey in the first place. At the same time, new words, deployed in the right way, can have the opposite effect as well. They can fascinate the reader enough to keep them glued to the page.

Writing for academic journals was difficult because I had to put on the voice of a type of idealized physicist that was informed by whiteness, maleness, and colonialism, yet was presented to me as simply objective. But it was also a given that even if my writing was bad, some physicists would still engage with it because we were trying to answer similar research questions. Now that I'm writing for the science curious—or to draw in someone who is actually reading an article right next to mine in the magazine—there is no minimal level of interest that I can take for granted. The academic voice that still exists within me is not very useful here. Often, what ends up being more helpful is remembering what learning physics felt like when I was a teenager.

In high school, learning new physics terms in class, or from popular science books, felt like being handed a key to a chest filled with new and precious knowledge of the universe. New words would stick in my mind like sesame seeds in my teeth, and the only way to dislodge them was to get as close as I could to knowing exactly what they meant. One of my physics teachers and I once tried to read a book about string theory and got stuck on what a "gauge" meant within the theory's mathematical formalism. I only fully grasped this in graduate school after I enrolled in a QFT course. I felt triumphant in that class—I had closed the loop on a word that had for years kept me excited for the day when I would finally learn QFT for real.

But what helped me understand gauges in that class was the working knowledge of vector spaces and mathematical operators that I'd gained from years of study in the meantime. If I had to use the phrase "gauge theory" in one of my articles today, I could not offer that sort of mathematics to my readers. A true statement that requires a college-level education in mathematics to understand will always fail to actually convey that truth to most people. Does its truth, then, really matter? I could try to communicate to readers the importance of gauges and convey my own excitement over learning about this new set of theories, but I could never use terms like "vector field" or "Lie group."

In this sense, I got my best lessons in science writing not by working with editors but by teaching ninth graders. My best classes and my best articles always had a lot in common. I was most successful when I could offer either the student, or the reader, a compelling hook to convince them to give me their time, as well as some hint of what the payoff would be. Once I've earned their investment, I couldn't give the student or the reader a chance to get bored or overwhelmed by information, and I couldn't make either of them feel like I'm speaking down to them. Learning physics is about building and rebuilding your mental image of how you think the world works in the most literal sense—and you are less likely to retool those images if you feel condescended to. It was easier for me to encourage this rebuilding process when I was in the classroom and constantly chatting with students, but on the page, the only assistance I can offer the reader is words.

The longer I write, the more malleable and versatile my words get. Instead of saying "wavefunction," I write "mathematical equation that captures the fact that all quantum objects are sometimes wave-like." Instead of "quasiparticle," I describe a new particle emerging from a collection of old ones, all of them moving in synchrony to form something new. Instead of "superposition of states," I write, "Here, an object has two options for what it can be like, but it is impossible to determine which one it actually chose, as if it occupies both at once."

When I get the urge to write "because of quantum effects" or "because the laws of physics say X," I try to remind myself that the reader wants to make sense of the world themselves, rather than just being told what it's like. I try to remind myself that they get to have a small "aha" moment, a chance to make their own connections. When I feel myself slipping into details that only a specialist looking to reproduce a study would need to know about, I try to remind myself that the reader's attention is precious. I make judgment calls, and I rely on a much broader array of words than I would have ever gotten away with in all those four-page, journal-ready papers. I spend a lot less time worrying whether I will betray the fact that a person, complete with affinities and quirks, decided how to put those words together.

Combined with the ever-present pressures of working at a for-profit magazine and having to play the search-engine-optimization game in order to generate web traffic, this process can be really scary. I worry constantly about misleading my readers and also misrepresenting the work of the (often young) researchers who've found the time to sit through an interview with me or write me long emails on short notice. I worry about no one seeing the story because I am naive about marketing, or the story getting distorted because my editors and I put some overly clever headline on top of it. I worry about disappointing the researcher in me who once tried so hard to write all of those extremely precise, official-sounding papers, and I worry about confusing someone like the teenage version of me, who loved to read the kind of writing that I produce now. Quite frankly, I worry about you, the person reading this book, and whether there is enough truth embedded in the fanciful and self-indulgent metaphors and analogies I have constructed in some of these pages.

But then I remember quantum spin, and how deeply wrong, yet deeply useful, that four-letter word is.

By 1927, five years after the Stern-Gerlach experiment, quantum mechanics had changed. Bohr's model of the atom had become "old quantum theory," and it was replaced by an updated and more accurate one. Bohr's ideas that carried over into new quantum theory were now also buttressed by a different mathematical foundation. Electrons, which had always been considered particles, were now being discussed by quantum physicists as waves, and, additionally, those physicists were using mathematical objects called wavefunctions and matrices in their calculations.[10] This version of quantum theory has stood the test of time well enough to be taught to me in college some eighty years later. During this overhaul, "space quantization" was discarded, and in its place, "spin"—the idea and the word—emerged and became a mainstay.

But the way "spin" got attached to the findings of Stern and Gerlach is roundabout and convoluted, underscoring how confused about the property everyone was—including those who named it. The first physicists to use the word in a published journal article, George Uhlenbeck and Samuel Goudsmit, were inspired not by experiments but by the mathematical arguments that Wolfgang Pauli had made about the atom, though without using the word.[11] Pauli himself was inspired by some experiments, but not Stern and Gerlach's.[12] Their silver atoms experiment, possibly because of its association with "space quantization," seems to have faded from view for several years, even though it now often gets cited as the moment when spin was discovered.

Within a few years of the Stern-Gerlach experiment, physicists determined that the lowest-energy electrons in silver atoms, like those in the experiment, would not experience magnetic forces in the way that they had if they were only moving in the orbits postulated by Bohr. The two marks on the screen that Gerlach had seen corresponded to two of the values that spin can have, which are 1/2 and -1/2, typically called "spin up" and "spin down." But that connection was only made by Ronald Fraser, in 1927, after he had read about the work of Uhlenbeck, Goudsmit, and Pauli.[13]

In a 1924 paper that got Uhlenbeck and Goudsmit thinking about particles spinning, Pauli used the phrase "classically non-describable two-valuedness of the quantum-theoretic properties of the light-electron" to refer to the quality that we now call "spin."[14] He stood by this phrase for most of his career. Certainly it is very detailed and very literal, and accurate to what he had discovered in the math. But, to me, it also raises more questions than it answers.

What does it mean for a physical property to be "classically non-describable"? Are there truly other ways we can describe it? And how can something be non-describable yet manifest itself in our classical lived experience, literally keeping matter from crumbling? What is this thing anyway? It is difficult to accept that something is simply non-describable, and then there's the question of whether that acceptance is useful.

The physicist Vlatko Vedral makes a case for not letting mathematics overshadow words and for using analogies despite their possible shortcomings. "Putting things in this mathematical way does not really illuminate us about what's going on physically when we do quantum experiments. To a physicist, mathematics is simply not enough. We'd like to understand quantum reality. Hence, analogies," he writes, in direct opposition to the kind of language that Pauli preferred.[15]

In their 1925 paper, Uhlenbeck and Goudsmit did not necessarily think that they were making an analogy, but they were certainly proposing something not just describable but that you could actually imagine. They appealed to the now familiar idea: if you take a charged particle and make it either rotate or spin, it will behave like a tiny magnet. Focusing on the electron, they associated one value of spin with the particle spinning clockwise, and the other with it spinning counterclockwise. This was much more elegant and easier to stomach than Pauli's "classically non-describable two-valuedness." It also did not take a lot of training in electromagnetism for most physicists to recognize it as qualitatively true.

Yet, a single calculation—one so simple as to now sometimes be assigned as a problem in college classes—almost immediately showed that what Uhlenbeck and Goudsmit were proposing was impossible. For an electron to spin fast enough to experience magnetic forces like those suggested by Pauli's work and by spectroscopy experiments, the electron would have to be spinning faster than the speed of light. In other words, it would have to break one of the most fundamental laws of the universe.

In fact, a year before Uhlenbeck and Goudsmit's paper, another young physicist, Ralph Kronig, had had the same idea about electrons spinning, but he didn't publish it because Pauli had privately torn it apart.[16] Uhlenbeck and Goudsmit also tried to retract their work, but were talked out of it by a mentor who told them they were young enough to "afford a stupidity."[17] Remarkably, their stupidity stuck. It even made it into textbooks.

I am telling you this long historical story, which features so many moments of discovery and so many scientists that are widely recognized as luminaries of their time, because at the center of it all is an ingredient of physical reality that, to this day, words cannot capture with complete accuracy. But physicists had to keep talking to each other about it, and they had to keep teaching their students, and if they wanted their research to be funded with the public's money, they had to find ways to talk about it to non-physicists too. Instead of prefacing every discussion of spin with a long and cumbersome intro about the mathematical structure of quantum theory, or sweeping the issue of how to think about spin under the rug by making up a new name for it, they did the more human thing and offered something that anyone who has ever seen a spinning top can hang their imagination onto. The more time I spend writing about physics professionally, the more I am convinced that this was actually a big success for the field.

There is hubris in thinking that we can ever perfectly understand the nature of physical reality, and even more hubris in thinking that we will have perfect words for expressing it. In our everyday conversations, we all use analogies and imperfectly precise phrases all the time. When I am speaking with a loved one, I often follow up some rambling, half-baked thought by saying, "But you know what I mean." I feel less comfortable doing that in my professional writing. This is in part because I was educated in a system of institutionalized science that taught me a lot about the value of precision and objectivity as a means of getting to the truth, but very little about how understanding something to be true also lies in feeling it. My high school students taught me this. They did their best problem-solving when the physics lessons I was teaching them found a way to merge with their intuition—when the lessons were not just rules to be memorized, but a more natural description of what the world is like.

Words do matter, and I really can't "just say things," but it is also my job to help words like "spin" leave the microcosm of physicists who are all in on the term's shortcomings and secrets, and into the world at large. I must offer those words a soft landing, regardless of whose lap they may fall into. In the process, those words and I constantly have to negotiate what it means for our shared efforts to be accurate, and to be true. And, as physicists do with "spin," I sometimes have to allow myself to use imperfect words to help make the nature of physical reality more perfectly felt by my readers.

I hope you know what I mean.

ACKNOWLEDGMENTS

It's a cliché to say this, but this is the book that I wish I could have read when I was younger. It is also the book that I wish I could have given to every brilliant person I've met over the years who told me that physics is not for them. I am immensely grateful to have had a chance to write it, and I can only hope that it will make you feel like physics is for you too, at least a little.

I want to thank every scientist who ever sat through an interview with me or sent me "quick comments in an email" when I was racing to meet some deadline, and especially those who have been receptive to the weird, speculative, broad-ranging questions that never made it into my news stories but certainly shaped some of the narratives in this book. I could never have had any of those loopy conversations had I not first been encouraged to think big and outside the box by my graduate school adviser, Smitha Vishveshwara, a physicist with the heart of a poet and an exceptional enjoyer of the world, whose guidance I am grateful for. She always attracted the absolute best people, and I shared both lots of camaraderie and lots of curiosity with Varsha Subramanyan and Suraj Hegde. Kathy Levin was the first person who made me feel like I could learn how to do theory and also fight my way to being taken seriously among physicists. I don't think I would have ever gone to graduate school had she not taken me under her wing. I am grateful to all my physics professors over the many, many years of my physics training, but especially Jeff Harvey, who taught me quantum mechanics out of Shankar, which always felt like a vote of confidence for my ability to handle advanced math. Rob Leigh and Eduardo Fradkin taught me quantum field theory, and Bryce Gadway taught me atomic physics and let me see ultracold atoms in his lab. Alex An and Shraddha Agrawal told me about what it was actually like to get those atoms to be so cold. None of my research into ultracold atoms would have ever happened without Kuei Sun and Courtney Lannert.

This book would not have existed without my work as a physics reporter at *New Scientist*, and I continue to be extremely grateful to have such an incredible position in the ever-precarious field of journalism. Everyone in the New York newsroom has supported this project from the jump, with kind words and care. Chelsea Whyte Slack-ed me many poems and strong feelings about particle nomenclature and the nature of time while I was writing this book, and I always loved her for that. Christie Taylor heard all my rants about the political nature of science and countered them with better ones—and we bonded over the queer undercurrents of so much of our respective work as science journalists. I would never have become a science journalist had Lee Billings, Kara Platoni, and Margaret Harris not given me a chance once, and I am immensely grateful for their willingness to work with a very green reporter and give them life-changing opportunities. The early-career science writers' group that I joined early in my New York days helped me stick with that life change. Pam Weintraub commissioned and edited a long piece that I wrote about imaginary numbers; William Callahan read it, offered me representation, and helped birth the idea for this book. Speaking with my editor, Catherine Tung, convinced me that I could write it; edits by Alison Rodriguez allowed me to complete it.

I am grateful to every student I ever interacted with at Bard High School Early College in Manhattan. They probably taught me as much as any of my professors. Thank you for showing me that science can be full of joy and wonder when I was at my most sad and scared.

My yoga teachers at Amara in Urbana-Champaign and Daya in Ridgewood helped me hate my body less and love its strength more. My boxing instructors at Rumble (Miles, Kory, and Chris) showed me that I can, in fact, always punch harder, jump higher, and lift heavier.

I am also thankful to the community of writers that I fell into through writing my newsletter, *Ultracold*, and especially Alicia Kennedy, who made efforts to keep those conversations going in her salons week to week.

Many of my friends have unwittingly contributed to this book simply by making my life better through their companionship. I probably would not have made it through college without Kaitlyn Lee, who continues to be a friend. In graduate school I organized with Brianne, Will, Angela, Gloria, Ryan, and many others. Irina went to the pool and the Turkish restaurant with me; Emily, Leo, Ingrid, and Nabil hung out with me at many soup nights, cake sales, and Wingspan game nights, and they all showed me how good a life you can have when you stop running away from yourself. Johanna deLeyer-Tiarks read the bulk of this book before anyone else, and her infinite enthusiasm and ardent love helped me keep writing it. The support of my best friend, Alex, my platonic soulmate and my younger sibling from a different universe, was immensely important for this work, and growing our connection as not just friends but also two writers over the past two years has been exhilarating. My mom was another belligerently passionate supporter of this project, and I could not be more grateful for that—I am glad to have rebuilt my relationship with you, Mom; it makes my world richer.

I don't really know how to properly thank my spouse, Bennett, because every day that I spend with him deserves its own page of saccharine notes about love, gratitude, and laughter, and I really don't have space for that. Thanks, champ, I love you.

NOTES

CHAPTER 1: MEMORY DEVICES

1. Interview, Sept. 2022. Quotes from interviews and casual conversations with physicists are left anonymous throughout this book, as this is not necessarily a story about one physics hero but more so about my experience as a small participant in the world of physics.

2. Donna Lu, "What Is a Quantum Computer?" *New Scientist*, Nov. 2, 2019, https://www.newscientist.com/question/what-is-a-quantum-computer/.

3. "How Quantum Computers Work: Explaining Qubits to Quantum Superposition," posted Feb. 23, 2023, by *New Scientist*, YouTube, 5:12, https://www.youtube.com/watch?v=WW7DKcrQ-7E&ab_channel=NewScientist.

4. Interview, Nov. 2024.

5. Karmela Padavic-Callaghan, "IBM Has Built the World's Largest Fridge for Quantum Computers," *New Scientist*, Sept. 8, 2022, https://www.newscientist.com/article/2337083-ibm-has-built-the-worlds-largest-fridge-for-quantum-computers/.

6. This section borrows from Karmela Padavic-Callaghan, "Convolution Reverb," *Ultracold* (newsletter), Mar. 27, 2022, https://ultracold.substack.com/p/convolution-reverb.

7. See chapters 5 and 10.

8. Nobel Prize in Physics 2001, https://www.nobelprize.org/prizes/physics/2001/summary/, accessed Nov. 26, 2024.

9. At the time, I was only using "she" and "her" pronouns.

10. Will Knight, "Entangled Photons Secure Money Transfer," *New Scientist*, Apr. 22, 2004, https://www.newscientist.com/article/dn4914-entangled-photons-secure-money-transfer/.

11. Alex Wilkins, "Quantum Internet Draws Near Thanks to Entangled Memory Breakthroughs," *New Scientist*, May 15, 2024, https://www.newscientist.com/article/2431464-quantum-internet-draws-near-thanks-to-entangled-memory-breakthroughs/; Meredith Fore, "Chicago Expands and Activates Quantum Network, Taking Steps Toward a Secure Quantum Internet," Chicago Quantum Exchange, June 16, 2022, https://chicagoquantum.org/news/chicago-expands-and

-activates-quantum-network-taking-steps-toward-secure-quantum-internet; Matt Swayne, "Qubitekk-Powered EPB Quantum Network Hosts Oak Ridge National Laboratory's First Run on a Commercial Quantum Network," *Quantum Insider*, Sept. 11, 2024, https://thequantuminsider.com/2024/09/11/qubitekk-powered-epb-quantum-network-hosts-oak-ridge-national-laboratorys-first-run-on-a-commercial-quantum-network/; Karmela Padavic-Callaghan, "An Unhackable Quantum Internet Is Being Built in New York City," *New Scientist*, Apr. 11, 2023, https://www.newscientist.com/article/2368353-an-unhackable-quantum-internet-is-being-built-in-new-york-city/.

12. "Consortium of Parties First in the World to Build Scalable Quantum Internet Connection in Rotterdam Port," Port of Rotterdam, May 14, 2024, https://www.portofrotterdam.com/en/news-and-press-releases/consortium-parties-first-world-build-scalable-quantum-internet-connection, accessed Nov. 25, 2024.

13. Jeremy Hsu, "Quantum Satellite Sets Globe-Spanning Distance Record," *New Scientist*, Mar. 19, 2025, https://www.newscientist.com/article/2472766-quantum-satellite-sets-globe-spanning-distance-record/.

14. Karmela Padavic-Callaghan, "The First Operating System for Quantum Networks Has Been Built," *New Scientist*, Mar. 12, 2025, https://www.newscientist.com/article/2471543-the-first-operating-system-for-quantum-networks-has-been-built/.

15. Karmela Padavic-Callaghan, "Ultra-Secure Quantum Data Sent over Existing Internet Cables," *New Scientist*, Apr. 23, 2025, https://www.newscientist.com/article/2477664-ultra-secure-quantum-data-sent-over-existing-internet-cables/.

16. Leah Crane, "Record-Breaking Quantum Memory Brings Quantum Internet One Step Closer," *New Scientist*, Feb. 12, 2020, https://www.newscientist com/article/2233317-record-breaking-quantum-memory-brings-quantum-internet-one-step-closer/.

17. Karmela Padavic-Callaghan, "Quantum City Simulation Shows How to Make Paris-Sized Quantum Internet," *New Scientist*, Nov. 21, 2022, https://www.newscientist.com/article/2347674-quantum-city-simulation-shows-how-to-make-paris-sized-quantum-internet/.

18. Matthew J. Sellars et al., "Optically Addressable Nuclear Spins in a Solid with a Six-Hour Coherence Time," *Nature* 517 (Jan. 2015): 177–80, https://doi.org/10.1038/nature14025.

19. H. P. Bartling et al., "Entanglement of Spin-Pair Qubits with Intrinsic Dephasing Times Exceeding a Minute," *Physical Review X* 12, no. 1 (Mar. 14, 2022): 011048, https://journals.aps.org/prx/abstract/10.1103/PhysRevX.12.011048.

20. Padavic-Callaghan, "An Unhackable Quantum Internet Is Being Built in New York City."

21. Karmela Padavic-Callaghan, "Quantum Memory Device Could Make Real-World Quantum Internet Possible," *New Scientist*, Sept. 20, 2022, https://

www.newscientist.com/article/2338849-quantum-memory-device-could-make-real-world-quantum-internet-possible/.

22. This section is adapted from Karmela Padavic-Callaghan, "Spectral Gap," *Ultracold* (newsletter), Aug. 6, 2020, https://ultracold.substack.com/p/spectral-gap.

23. Karmela Padavic-Callaghan, "Hollow Condensates, Topological Ladders and Quasiperiodic Chains," PhD diss., University of Illinois at Urbana-Champaign, 2020.

CHAPTER 2: INDEFINITE CAUSALITY

1. Giulio Chiribella et al., "Quantum Computations Without Definite Causal Structure," *Physical Review A* 88, no. 2 (Aug. 2013): 022318, https://journals.aps.org/pra/abstract/10.1103/PhysRevA.88.022318.

2. Ognyan Oreshkov, Fabio Costa, and Časlav Brukner, "Quantum Correlations with No Causal Order," *Nature Communications* 3 (Oct. 2012): 1092, https://www.nature.com/articles/ncomms2076.

3. Karmela Padavic-Callaghan, "Self-Trapping," *Ultracold* (newsletter), Jan. 17, 2020, https://ultracold.substack.com/p/self-trapping.

4. Karmela Padavic-Callaghan, "Quantum Batteries Could Charge Better by Breaking Rules of Causality," *New Scientist*, Nov. 3, 2023, https://www.newscientist.com/article/2401198-quantum-batteries-could-charge-better-by-breaking-rules-of-causality/.

5. Chiribella et al., "Quantum Computations Without Definite Causal Structure."

6. Giulia Rubino et al., "Experimental Verification of an Indefinite Causal Order," *Science Advances* 3, no. 3 (Mar. 2017): e1602589, https://www.science.org/doi/10.1126/sciadv.1602589.

7. Natalie Wolchover, "Quantum Mischief Rewrites the Laws of Cause and Effect," *Quanta Magazine*, Mar. 11, 2021, https://www.quantamagazine.org/quantum-mischief-rewrites-the-laws-of-cause-and-effect-20210311/.

8. Interview, May 2024.

9. This section borrows from Padavic-Callaghan, "Self-Trapping."

10. Kejin Wei et al., "Experimental Quantum Switching for Exponentially Superior Quantum Communication Complexity," *Physical Review Letters* 122, no. 12 (Mar. 2019): 120504, https://journals.aps.org/prl/abstract/10.1103/PhysRevLett.122.120504.

11. Michael Brooks, "How Indefinite Causality Could Lead Us to a Theory of Quantum Gravity," *New Scientist*, May 20, 2024, https://www.newscientist.com/article/mg26234921-100-how-indefinite-causality-could-lead-us-to-a-theory-of-quantum-gravity/.

12. Karmela Padavic-Callaghan, "Cause and Effect May Not Actually Be Muddled in the Quantum Realm," *New Scientist*, Sept. 13, 2024, https://www.newscientist.com/article/2447251-cause-and-effect-may-not-actually-be-muddled-in-the-quantum-realm/.

CHAPTER 3: PINK NOISE AND THE ARROW OF TIME

1. Lee Billings, "2 Futures Can Explain Time's Mysterious Past," *Scientific American*, Dec. 8, 2014, https://www.scientificamerican.com/article/2-futures-can-explain-time-s-mysterious-past/.

2. Arthur S. Eddington, *The Nature of the Physical World* (Cambridge: Cambridge University Press, 1948), 35.

3. Alfredo González-Espinoza, Gustavo Martínez-Mekler, and Lucas Lacasa, "Arrow of Time Across Five Centuries of Classical Music," *Physical Review Research* 2, no. 3 (July 2020): 033166, https://journals.aps.org/prresearch/abstract/10.1103/PhysRevResearch.2.033166.

4. This section is adapted from Karmela Padavic-Callaghan, "Pink Noise," *Ultracold* (newsletter), Aug. 19, 2021, https://ultracold.substack.com/p/pink-noise.

5. AC/DC, "The Jack," track 3, *T.N.T.*, Atlantic Records, 1976.

6. Sarah Elizabeth Culpepper, "Musical Time and Information Theory Entropy," master's thesis, University of Iowa, 2010, p. 74, https://iro.uiowa.edu/esploro/outputs/graduate/Musical-time-and-information-theory-entropy/9983776837602771/filesAndLinks?index=0.

7. Martin Scherzinger with Neville Hoad, "Anton Webern and the Concept of Symmetrical Inversion: A Reconsideration on the Terrain of Gender," *repercussions* 6, no. 2 (1997): 63–147.

8. Harry Sword, *Monolithic Undertow: In Search of Sonic Oblivion* (London: White Rabbit, 2021), 146.

9. Sword, *Monolithic Undertow*, 155.

10. Sword, *Monolithic Undertow*, 155.

11. Interview, Aug. 2020.

12. Karmela Padavic-Callaghan, "Time's Arrow Flies Through 500 Years of Classical Music, Physicists Say," *Scientific American*, Aug. 19, 2020, https://www.scientificamerican.com/article/times-arrow-flies-through-500-years-of-classical-music-physicists-say/.

13. Richard C. Pinkerton, "Information Theory and Melody," *Scientific American*, Feb. 1956, 77–87.

14. Queen, "Innuendo," track 1, *Innuendo*, Parlophone, 1991.

15. González-Espinoza, Martínez-Mekler, and Lacasa, "Arrow of Time Across Five Centuries of Classical Music," 033166.

16. Gonçalo Marquesa et al., "Short-Term Feature Space and Music Genre Classification," *Journal of New Music Research* 40, no. 2 (June 2011): 127–37.

17. Iannis Xenakis, *Formalized Music: Thought and Mathematics in Composition* (Hillsdale, NY: Pendragon Press, 1992), 255.

18. Karmela Padavic-Callaghan, "Mathematicians Have Finally Proved That Bach Was a Great Composer," *New Scientist*, Feb. 2, 2024, https://www.newscientist.com/article/2415469-mathematicians-have-finally-proved-that-bach-was-a-great-composer/.

19. Interview, Feb. 2024.

20. "Queen (Live at Wembley Stadium 1986)," posted Mar. 26, 2021, by Top Music Rock, YouTube, 1 hr. 49 min., 38 sec., https://www.youtube.com/watch?v=p9a9ZVTACCY&ab_channel=Topmusicrock.

21. This section is adapted from Padavic-Callaghan, "Pink Noise."

22. Natalie Wolchover, "Time's Arrow Traced to Quantum Source," *Quanta Magazine*, Apr. 2014, https://www.quantamagazine.org/quantum-entanglement-drives-the-arrow-of-time-scientists-say-20140416/.

23. Wolchover, "Time's Arrow Traced to Quantum Source."

24. Liam Graham, *Molecular Storms: The Physics of Stars, Cells and the Origin of Life* (Cham, Switzerland: Springer, 2023), 45.

CHAPTER 4: THE MANY-WORLDS INTERPRETATION

1. Peter Byrne, "The Many Worlds of Hugh Everett," *Scientific American*, Oct. 21, 2008, https://www.scientificamerican.com/article/hugh-everett-biography/.

2. Werner Heisenberg, *Physics and Philosophy: The Revolution in Modern Science* (London: Penguin Classics, 2000), 22.

3. Claus Kiefer, "On the Interpretation of Quantum Theory—from Copenhagen to the Present Day," *arXiv*, Oct. 22, 2002, https://arxiv.org/abs/quant-ph/0210152; Werner Heisenberg, "Über den anschaulichen Inhalt der quantentheoretischen Kinematik und Mechanik," *Zeitschrift fur Physik* 43 (Mar. 1927): 172–98, https://doi.org/10.1007/BF01397280.

4. Karmela Padavic-Callaghan, "In a First, Physicists Glimpse a Quantum Ghost," *Scientific American*, Dec. 8, 2021, https://www.scientificamerican.com/article/in-a-first-physicists-glimpse-a-quantum-ghost/.

5. Gerard 't Hooft et al., "The Sounds of Science—a Symphony for Many Instruments and Voices: Part II," *Physica Scripta* 99, no. 5 (Apr. 2024): 052501, https://iopscience.iop.org/article/10.1088/1402-4896/ad2abe.

6. Karmela Padavic-Callaghan, "Can We Solve Quantum Theory's Biggest Problem by Redefining Reality?" *New Scientist*, Sept. 4, 2024, https://www.newscientist.com/article/mg26335070-700-can-we-solve-quantum-theorys-biggest-problem-by-redefining-reality/.

7. Padavic-Callaghan, "Can We Solve Quantum Theory's Biggest Problem by Redefining Reality?" 25.

8. Sean Carroll, "Splitting the Universe," *Aeon Magazine*, Sept. 11, 2019, https://aeon.co/essays/how-the-many-worlds-theory-of-hugh-everett-split-the-universe.

9. Hugh Everett III, "'Relative State' Formulation of Quantum Mechanics," *Reviews of Modern Physics* 29, no. 3 (July 1957): 454–62, https://journals.aps.org/rmp/abstract/10.1103/RevModPhys.29.454.

10. Emily Qureshi-Hurst, "Many Worlds, Many Selves," *Aeon*, Nov. 28, 2024, https://aeon.co/essays/why-identity-morality-and-faith-splinter-in-the-multiverse?.

11. Philip Ball, "Why the Many-Worlds Interpretation Has Many Problems," *Quanta Magazine*, Oct. 18, 2018, https://www.quantamagazine.org/why-the-many-worlds-interpretation-has-many-problems-20181018/.

12. Carroll, "Splitting the Universe."

13. Rowan Hooper, "Hugh Everett: The Man Who Gave Us the Multiverse," *New Scientist*, Sept. 24, 2014, https://www.newscientist.com/article/dn26261-hugh-everett-the-man-who-gave-us-the-multiverse/.

14. Bryce S. DeWitt, "Quantum Mechanics and Reality," *Physics Today* 23, no. 9 (Sept. 1970): 30–35, http://dx.doi.org/10.1063/1.3022331.

15. Wojciech H. Zurek, "Decoherence and the Transition from Quantum to Classical—REVISITED," *arXiv*, June 10, 2003, https://arxiv.org/abs/quant-ph/0306072.

16. Carroll, "Splitting the Universe."

17. Karmela Padavic-Callaghan, "The Multiverse Could Be Much, Much Bigger Than We Ever Imagined," *New Scientist*, Apr. 9, 2024, https://www.newscientist.com/article/mg26234860-100-the-multiverse-could-be-much-much-bigger-than-we-ever-imagined/.

18. Interview, Apr. 2024.

19. Eugene B. Shikhovtsev (author) and Kenneth W. Ford (editor), "Biographical Sketch of Hugh Everett, III," MIT Kavli Institute, https://space.mit.edu/home/tegmark/everett/everett.html, accessed Nov. 24, 2024.

20. Rowan Hooper, "Multiverse Me: Should I Care About My Other Selves?" *New Scientist*, Sept. 24, 2014, https://www.newscientist.com/article/mg22329880-400-multiverse-me-should-i-care-about-my-other-selves/.

21. Hooper, "Multiverse Me."

22. Qureshi-Hurst, "Many Worlds, Many Selves."

23. Qureshi-Hurst, "Many Worlds, Many Selves."

24. Qureshi-Hurst, "Many Worlds, Many Selves."

CHAPTER 5: SCHRÖDINGER'S CAT

1. Smitha Vishveshwara and Latrelle Bright, *Quantum Voyages*, play script, 2018.

2. John D. Trimmer, "The Present Situation in Quantum Mechanics: A Translation of Schrödinger's 'Cat Paradox' Paper," *Proceedings of the American Philosophical Society* 124, no. 5 (Oct. 1980): 323–38, https://www.jstor.org/stable/986572.

3. Trimmer, "The Present Situation in Quantum Mechanics."

4. Trimmer, "The Present Situation in Quantum Mechanics."

5. William Gibson, *Pattern Recognition: A Novel* (New York: Berkley, 2004), 5.

6. Louisa Gilder, *The Age of Entanglement: When Quantum Physics Was Reborn* (New York: Alfred A. Knopf, 2008), 82–85.

7. Peter Debye, "Peter Debye–Session III," interview by Thomas S. Kuhn and George Uhlenbeck, American Institute of Physics, May 4, 1962, https://repository.aip.org/debye-peter-j-w-1962-may-3-and-4-session-iii.

8. Gilder, *The Age of Entanglement*, 84.

9. Gilder, *The Age of Entanglement*, 85.

10. Gilder, *The Age of Entanglement*, 84.

11. Hugh Everett III, "'Relative State' Formulation of Quantum Mechanics," *Reviews of Modern Physics* 29, no. 3 (July 1957): 454–62, https://journals.aps.org/rmp/abstract/10.1103/RevModPhys.29.454, and see previous chapter; Max Born, "The Statistical Interpretations of Quantum Mechanics," Nobel Prize lecture, Dec. 11, 1954, https://www.nobelprize.org/prizes/physics/1954/born/lecture/.

12. Gilder, *The Age of Entanglement*, 176.

13. Ramamurti Shankar, *Principles of Quantum Mechanics*, 2nd ed. (New York: Plenum Press, 1994), 115–51.

14. Shankar, *Principles of Quantum Mechanics*, 115–16.

15. Trimmer, "The Present Situation in Quantum Mechanics," 323–38.

16. "Physics Education Statistics," American Physical Society, https://www.aps.org/learning-center/statistics/education, accessed Nov. 30, 2024.

17. "Mapping Attacks on LGBTQ Rights in U.S. State Legislatures in 2024," ACLU, https://www.aclu.org/legislative-attacks-on-lgbtq-rights-2024, accessed Nov. 21, 2024; Priya Krishnakumar, "This Record-Breaking Year for Anti-transgender Legislation Would Affect Minors the Most," CNN, Apr. 15, 2021, https://www.cnn.com/2021/04/15/politics/anti-transgender-legislation-2021/index.html.

18. Mark Plavetich, "Croatian American Community in New York and the American Spirit–Part 1," Croatians Online, Apr. 2, 2024, https://croatiansonline.com/en/croatian-american-community-in-new-york-and-the-american-spirit-part-1/.

19. Sabine Hossenfelder, "Why Is Quantum Mechanics Weird? The Bomb Experiment," *Backreaction* (blog), Aug. 28, 2021, https://backreaction.blogspot.com/2021/08/why-is-quantum-mechanics-weird-bomb.html.

20. Leah Crane, "A Macroscopic Amount of Matter Has Been Put in a Quantum Superposition," *New Scientist*, Apr. 12, 2023, https://www.newscientist.com/article/2368306-a-macroscopic-amount-of-matter-has-been-put-in-a-quantum-superposition/.

21. Karmela Padavic-Callaghan, "Quantum Computer Uses a Time Crystal as a Control Dial," *New Scientist*, Feb. 7, 2024, https://www.newscientist.com/article/2415611-quantum-computer-uses-a-time-crystal-as-a-control-dial/.

22. N. David Mermin, "Quantum Mysteries Revisited," *American Journal of Physics* 58, no. 8 (Aug. 1990): 731–34, https://doi.org/10.1119/1.16503.

23. Padavic-Callaghan, "Quantum Computer Uses a Time Crystal as a Control Dial."

24. Email, Feb. 2024.

25. Mario Krenn, "Largest Genuine Entanglement: Qubits in GHZ State," updated May 8, 2025, https://mariokrenn.wordpress.com/2021/01/29/reference

-list-for-records-in-large-entanglement-generation-number-of-qubits-in-ghz-states/.

26. Karmela Padavic-Callaghan, "How Schrödinger's Cat Could Make Quantum Computers Work Better," *New Scientist*, May 6, 2024, https://www.newscientist.com/article/2429855-how-schrodingers-cat-could-make-quantum-computers-work-better/.

27. Cate Lawrence, "Alice & Bob and Partners Awarded €16.5M to Slash Quantum Computing Costs," *Tech.eu*, Mar. 26, 2024, https://tech.eu/2024/03/26/alice-bob-and-partners-awarded-eur165m-to-slash-quantum-computing-costs.

28. Yakir Aharonov, David Z. Albert, and Lev Vaidman, "How the Result of a Measurement of a Component of the Spin of a Spin-1/2 Particle Can Turn Out to Be 100," *Physical Review Letters* 60, no. 14 (Apr. 1988): 1351–54, https://journals.aps.org/prl/abstract/10.1103/PhysRevLett.60.1351; Sacha Kocsis et al., "Observing the Average Trajectories of Single Photons in a Two-Slit Interferometer," *Science* 332, no. 6034 (June 2011): 1170–73, https://www.science.org/doi/10.1126/science.1202218; Aephraim Steinberg et al., "In Praise of Weakness," *Physics World* 26, no. 3 (Mar. 2013): 47, https://iopscience.iop.org/article/10.1088/2058-7058/26/03/36.

29. Lev Vaidman, "Weak Value Controversy," *Philosophical Transactions of the Royal Society A* 375 (Oct. 2017): 20160395, https://royalsocietypublishing.org/doi/10.1098/rsta.2016.0395.

30. Sam Levin, "Ice Agents Detain US Citizen as LA Immigration Raids Continue: 'It's Racial Profiling,'" *The Guardian*, June 16, 2025, https://www.theguardian.com/us-news/2025/jun/16/los-angeles-immigration-raids-montebello.

CHAPTER 6: PATH INTEGRAL

1. Richard P. Feynman, "Space-Time Approach to Non-relativistic Quantum Mechanics," *Reviews of Modern Physics* 20 (Apr. 1948): 367–87, https://doi.org/10.1103/RevModPhys.20.367.

2. "The Best Mind Since Einstein," dir. Christopher Sykes, *Nova*, WGBH Boston, 2006.

3. Feynman, "Space-Time Approach to Non-relativistic Quantum Mechanics," 367–87.

4. P. A. M. Dirac, "The Lagrangian in Quantum Mechanics," *Physikalische Zeitschrift der Sowjetunion* 3, no. 1 (Nov. 1933): 64–72, https://www.worldscientific.com/doi/10.1142/9789812567635_0003; N. D. Hari Dass, "Dirac and the Path Integral," *arXiv*, Mar. 28, 2020, https://arxiv.org/abs/2003.12683.

5. Feynman, "Space-Time Approach to Non-relativistic Quantum Mechanics," 367–87.

6. "Richard Feynman—the Character of Physical Law—Part 6 Probability and Uncertainty (Full Version)," posted Jan. 30, 2011, by Tinkerin' Thinkers, YouTube, 56 min., 9 sec., https://www.youtube.com/watch?v=aAgcqgDc-YM&t=483s&ab_channel=Tinkerin%27Thinkers.

7. Web of Stories—Life Stories of Remarkable People, "Freeman Dyson—Talking Physics with Feynman: Path Integrals (71/157)," Sept. 7, 2016, YouTube, 2:55, https://www.youtube.com/watch?v=1ZDKGFNOzXc&ab_channel=WebofStories-LifeStoriesofRemarkablePeople.

8. Leila McNeill, "Surely You're a Creep, Mr. Feynman," *The Baffler*, Jan. 2019, https://thebaffler.com/outbursts/surely-youre-a-creep-mr-feynman-mcneill; Shannon Palus, "Famed Physicist Richard Feynman Was Known as an Odd Genius. Was He Also an Abuser?" *Slate*, Jan. 14, 2019, https://slate.com/technology/2019/01/richard-feynman-physical-abuse-science-wife-fbi.html; Richard P. Feynman, "What Is Science?" *Physics Teacher* 7, no. 6 (Sept. 1969): 313–20, https://pubs.aip.org/aapt/pte/article-abstract/7/6/313/278346/What-Is-Science?

9. Sergio Albeverio and Sonia Mazzucchi, "Path Integral: Mathematical Aspects," *Scholarpedia* 6, no. 1 (2011): 8832, http://www.scholarpedia.org/article/Path_integral:_mathematical_aspects.

10. Michael Murray, Alan Caret, and Peter Bouwknegt, "Millennium Prize: The Yang-Mills Existence and Mass Gap problem," *The Conversation*, Dec. 8, 2011, https://theconversation.com/millennium-prize-the-yang-mills-existence-and-mass-gap-problem-3848.

11. Peter Woit, "The Trouble with Path Integrals, Part I," *Not Even Wrong* (blog), Feb. 16, 2023, https://www.math.columbia.edu/~woit/wordpress/?p=13319.

12. Charlie Wood, "How Our Reality May Be a Sum of All Possible Realities," *Quanta Magazine*, Feb. 6, 2023, https://www.quantamagazine.org/how-our-reality-may-be-a-sum-of-all-possible-realities-20230206/.

13. Wood, "How Our Reality May Be a Sum of All Possible Realities."

14. Patrick J. Mulvey, "Where Do New PhDs Work?" *Physics Today* 73, no. 10 (Oct. 2020): 40–46, https://pubs.aip.org/physicstoday/article/73/10/40/853141/Where-do-new-PhDs-work-Survey-questionnaire-data.

15. Ramón S. Barthelemy et al., "LGBT+ Physicists: Harassment, Persistence, and Uneven Support," *Physical Review Physics Education Research* 18, no. 1 (Mar. 2022): 010124, https://journals.aps.org/prper/abstract/10.1103/PhysRevPhysEducRes.18.010124.

16. Kiran G. L. Lee et al., "Meta-Research: The Effect of the COVID-19 Pandemic on the Gender Gap in Research Productivity Within Academia," *eLife* 12 (July 2023): e85427, https://elifesciences.org/articles/85427.

17. Celeste Suart, Kaitlyn Neuman, and Ray Truant, "The Impact of the COVID-19 Pandemic on Perceived Publication Pressure Among Academic Researchers in Canada," *PloS ONE* 17, no. 6 (June 2022): e0269743, https://doi.org/10.1371/journal.pone.0269743.

18. Janna Koretz, "What Happens When Your Career Becomes Your Whole Identity," *Harvard Business Review*, Dec. 26, 2019, https://hbr.org/2019/12/what-happens-when-your-career-becomes-your-whole-identity.

19. Kate Morgan, "Why We Define Ourselves by Our Jobs," BBC, Apr. 13, 2021, https://www.bbc.com/worklife/article/20210409-why-we-define-ourselves-by-our-jobs.

20. Chanda Prescod-Weinstein, *The Disordered Cosmos: A Journey into Dark Matter, Spacetime, and Dreams Deferred* (New York: Bold Type Books, 2022), 87.

21. Prescod-Weinstein, *The Disordered Cosmos*, 229.

22. William E. Sedlacek and Ryan D. Duffy, "The Presence of and Search for a Calling: Connections to Career Development," *Journal of Vocational Behavior* 70, no. 3 (June 2007): 590–601, https://doi.org/10.1016/j.jvb.2007.03.007.

23. Karl Marx, *Economic and Philosophic Manuscripts of 1844* (Moscow: Progress Publishers, 1959), https://www.marxists.org/archive/marx/works/1844/manuscripts/preface.htm.

24. Bruce Poulsen, "On the End of History Illusion," *Psychology Today*, Jan. 27, 2013, https://www.psychologytoday.com/us/blog/reality-play/201301/the-end-history-illusion.

25. Wood, "How Our Reality May Be a Sum of All Possible Realities."

26. Wood, "How Our Reality May Be a Sum of All Possible Realities."

CHAPTER 7: RIGID BODY DYNAMICS

1. William Thomson, "On Vortex Atoms," *Proceedings of the Royal Society of Edinburgh VI* (1867): 94–105, https://zapatopi.net/kelvin/papers/on_vortex_atoms.html.

2. Leah Crane, "Knot Theory Could Help Spacecraft Navigate Crowded Solar Systems," *New Scientist*, Apr. 19, 2024, https://www.newscientist.com/article/2427669-knot-theory-could-help-spacecraft-navigate-crowded-solar-systems/; Karmela Padavic-Callaghan, "Quantum Money That Uses the Mathematics of Knots Could Be Unforgeable," *New Scientist*, Jan. 7, 2023, https://www.newscientist.com/article/2352652-quantum-money-that-uses-the-mathematics-of-knots-could-be-unforgeable/.

3. This section borrows from Karmela Padavic-Callaghan, "Knot Theory," *Ultracold* (newsletter), July 20, 2017, https://ultracold.substack.com/p/knot-theory.

4. Larissa Fedunik, "How Many Knots Exist? A New Computing Trick Is Untangling the Answer," *New Scientist*, July 12, 2022, https://www.newscientist.com/article/mg25533950-900-how-many-knots-exist-a-new-computing-trick-is-untangling-the-answer/.

5. Karmela Padavic-Callaghan, "Blobs of Worms Untangle in Milliseconds with a Corkscrew Wiggle," *New Scientist*, Apr. 27, 2023, https://www.newscientist.com/article/2370982-blobs-of-worms-untangle-in-milliseconds-with-a-corkscrew-wiggle/.

6. Interview, Apr. 2023.

7. Joan Licata, "What the Mathematics of Knots Reveals About the Shape of the Universe," *New Scientist*, Jan. 5, 2024, https://www.newscientist.com/article/2410483-what-the-mathematics-of-knots-reveals-about-the-shape-of-the-universe/.

8. James O'Donnell and Casey Crownhart, "We Did the Math on AI's Energy Footprint. Here's the Story You Haven't Heard," *Technology Review*, May 20, 2025, https://www.technologyreview.com/2025/05/20/1116327/ai-energy-usage-climate-footprint-big-tech/.

9. Matthew Sparkes, "DeepMind AI Collaborates with Humans on Two Mathematical Breakthroughs," *New Scientist*, Dec. 1, 2021, https://www.newscientist.com/article/2299564-deepmind-ai-collaborates-with-humans-on-two-mathematical-breakthroughs/.

10. This section borrows from Karmela Padavic-Callaghan, "Isomerization," *Ultracold* (newsletter), Dec. 16, 2021, https://ultracold.substack.com/p/isomerization.

11. Jia Tolentino, *Trick Mirror: Reflections on Self-Delusion* (New York: Random House, 2020), 63–95.

12. Tolentino, *Trick Mirror*, 81.

13. Tolentino, *Trick Mirror*, 76.

14. Karmela Padavic-Callaghan, "Hollow Condensates, Topological Ladders and Quasiperiodic Chains," PhD diss., University of Illinois at Urbana-Champaign, 2020.

15. Padavic-Callaghan, "Knot Theory."

16. Padavic-Callaghan, "Knot Theory"; Karmela Padavic-Callaghan, "Compactified Dimensions," *Ultracold* (newsletter), Sept. 10, 2018, https://ultracold.substack.com/p/compactified-dimensions.

17. Robert M. Wald, *General Relativity* (Chicago: University of Chicago Press, 1984), 4.

18. Karmela Padavic-Callaghan, "Einstein's Theories Tested on the Largest Scale Ever—He Was Right," *New Scientist*, Nov. 20, 2024, https://www.newscientist.com/article/2456766-einsteins-theories-tested-on-the-largest-scale-ever-he-was-right/.

CHAPTER 8: ULTRACOLD

1. Karmela Padavic-Callaghan, "The Coolest Physics You've Ever Heard Of," *Scientific American*, Jan. 20, 2020, https://www.scientificamerican.com/blog/observations/the-coolest-physics-youve-ever-heard-of/.

2. Charles Seife, "Just Cool It—Atom-Chillers Win the Ultimate Accolade," *New Scientist*, Oct. 25, 1997, https://www.newscientist.com/article/mg15621052-200-just-cool-it-atom-chillers-win-the-ultimate-accolade/.

3. "A Brief History of Temperature," IOP Resources, https://spark.iop.org/brief-history-temperature, accessed Nov. 18, 2024.

4. "The Nobel Prize in Physics 2001," Nobel Prize, https://www.nobelprize.org/prizes/physics/2001/summary/, accessed Nov. 18, 2024.

5. William D. Phillips, "Laser Cooling and Trapping of Neutral Atoms," Nobel lecture, Dec. 8, 1997, https://www.nobelprize.org/prizes/physics/1997/phillips/lecture/.

6. Karmela Padavic-Callaghan, "The Strange Physics of Absolute Zero and What It Takes to Get There," *New Scientist*, Dec. 14, 2022, https://www.newscientist.com/article/mg25634171-900-the-strange-physics-of-absolute-zero-and-what-it-takes-to-get-there/.

7. Karmela Padavic-Callaghan, "Quantum Magnet Is Billions of Times Colder Than Interstellar Space," *New Scientist*, Sept. 1, 2022, https://www.newscientist.com/article/2336247-quantum-magnet-is-billions-of-times-colder-than-interstellar-space/.

8. Interview, Aug. 2022.

9. Kim Parker, Juliana Menasce Horowitz, and Renee Stepler, "On Gender Differences, No Consensus on Nature vs. Nurture," Pew Research Center, Dec. 5, 2017, https://www.pewresearch.org/social-trends/2017/12/05/on-gender-differences-no-consensus-on-nature-vs-nurture/.

10. David Olmos, "When Watching Others in Pain, Women's Brains Show More Empathy," UCLA Newsroom, Feb. 27, 2019, https://newsroom.ucla.edu/stories/womens-brains-show-more-empathy.

11. Brennan McDonald and Philipp Kanske, "Gender Differences in Empathy, Compassion, and Prosocial Donations, but Not Theory of Mind in a Naturalistic Social Task," *Scientific Reports* 13 (Nov. 2023): 20748, https://www.nature.com/articles/s41598-023-47747-9.

12. Karmela Padavic-Callaghan, "Ultracold Molecule Mystery Solved," *Scientific American*, Dec. 1, 2020, https://www.scientificamerican.com/article/ultracold-molecule-mystery-solved/.

13. Susan T. Fiske et al., "A Model of (Often Mixed) Stereotype Content: Competence and Warmth Respectively Follow from Perceived Status and Competition," *Journal of Personality and Social Psychology* 82, no. 6 (2002): 878–902, https://doi.org/10.1037/0022-3514.82.6.878.

14. Susan T. Fiske, "Stereotype Content: Warmth and Competence Endure," *Current Directions in Psychological Science* 27, no. 2 (Apr. 2018): 67–73, https://doi.org/10.1177/0963721417738825.

15. Fiske, "Stereotype Content."

16. Victoria Masterson, "Women Are Burning Out Doing Invisible 'Office Housework,'" World Economic Forum, Nov. 8, 2021, https://www.weforum.org/stories/2021/11/women-workplace-2021-invisible-labour/; Alexis Krivkovich et al., *Women in the Workplace 2024: The 10th-Anniversary Report*, McKinsey & Company, Sept. 17, 2024, https://www.mckinsey.com/featured-insights/diversity-and-inclusion/women-in-the-workplace#/; Joan C. Williams et al., "Climate

Control: Gender and Racial Bias in Engineering?" Center for WorkLife Law and Society of Women Engineers, 2016.

17. Elle Hunt, "'No Reward or Recognition': Why Women Should Say No to 'Office Housework,'" *The Guardian*, May 9, 2022, https://www.theguardian.com/society/2022/may/09/they-feel-guilty-why-women-should-say-no-to-office-housework.

18. Chanda Prescod-Weinstein, *The Disordered Cosmos: A Journey into Dark Matter, Spacetime, and Dreams Deferred* (New York: Bold Type Books, 2022), 187.

19. Grace Lee Boggs and Scott Kurashige, *The Next American Revolution: Sustainable Activism for the Twenty-First Century* (Oakland: University of California Press, 2012), 50.

20. Karmela Padavic-Callaghan, "Atoms at Temperatures Beyond Absolute Zero May Be a New Form of Matter," *New Scientist*, June 7, 2024, https://www.newscientist.com/article/2434069-atoms-at-temperatures-beyond-absolute-zero-may-be-a-new-form-of-matter/.

CHAPTER 9: DIFFRACTION AND SPECTROSCOPY

1. Taylor Swift, "Vigilante Shit," track 8, *Midnights*, Republic Records, 2022.

2. Jonathan J. P. Peters et al., "Event-Responsive Scanning Transmission Electron Microscopy," *Science* 385, no. 6708 (Aug. 2024): 549–53, https://www.science.org/doi/10.1126/science.ado8579.

3. Email, July 29, 2024.

4. Peter W. Hawkes, "Ernst Ruska," *Physics Today* 43, no. 7 (July 1990): 84–85, https://pubs.aip.org/physicstoday/article/43/7/84/406199/Ernst-Ruska.

5. Arabelle Sicardi, "Makeup's Dirty Little Secret: Covering the Scars of Abuse," *Jezebel*, Sept. 23, 2014, https://www.jezebel.com/makeups-dirty-little-secret-covering-the-scars-of-abuse.

6. Sicardi, "Makeup's Dirty Little Secret."

7. Raechel Anne Jolie, "style as a form of refusal," *radical love letters* (newsletter), Feb. 26, 2024, https://raechelannejolie.substack.com/p/style-as-a-form-of-refusal.

8. Nam Mai Phan et al., "Interaction of Fixed Number of Photons with Retinal Rod Cells," *Physical Review Letters* 112, no. 21 (May 2014): 213601, https://journals.aps.org/prl/abstract/10.1103/PhysRevLett.112.213601.

9. Jonathan N. Tinsley et al., "Direct Detection of a Single Photon by Humans," *Nature Communications* 7 (Jan. 2016): 12172, https://doi.org/10.1038/ncomms12172.

10. Anil Ananthaswamy, "The Human Eye Could Help Test Quantum Mechanics," *Scientific American*, July 10, 2018, https://www.scientificamerican.com/article/the-human-eye-could-help-test-quantum-mechanics/.

11. R. D. Archer-Hind, ed., *The Timaeus of Plato* (London: Macmillan and Co., 1888), 153.

12. Peter Adamson, *Philosophy in the Islamic World: A History of Philosophy Without Any Gaps* (Oxford: Oxford University Press, 2016), 77–83.

13. Karmela Padavic-Callaghan, various videos, https://www.tiktok.com/@kpcallaghan.

14. Ellyn Briggs, "Gen Zers Still Really Want to Be Influencers," *Morning Consult*, Oct. 4, 2023, https://pro.morningconsult.com/analysis/gen-z-interest-influencer-marketing.

15. Emily Sundberg, "The Machine in the Garden," *Feed Me* (newsletter), Aug. 9, 2024, https://www.readfeedme.com/p/the-machine-in-the-garden.

16. Margeaux Feldman (@softcore_trauma), various posts, https://www.instagram.com/softcore_trauma/?hl=en.

17. Margeaux Feldman (@softcore_trauma), "let's face it, we're gonna trigger each other. and if we're not, we're missing something," July 22, 2024, https://www.instagram.com/softcore_trauma/p/C9vgt3byaSj/?hl=en&img_index=1.

18. Margeaux Feldman, "Let's Face It: We're Gonna Trigger Each Other," *Carescapes* (newsletter), July 22, 2024, https://carescapes.substack.com/p/lets-face-it-were-gonna-trigger-each.

CHAPTER 10: WAVE-PARTICLE DUALITY

1. Richard K. Gehrenbeck, "Electron Diffraction: Fifty Years Ago," *Physics Today* 31, no. 1 (Jan. 1978): 34–41, https://doi.org/10.1063/1.3001830.

2. Clinton Davisson and Charles H. Kunsman, "The Scattering of Low Speed Electrons by Platinum and Magnesium," *Physical Review* 22, no. 3 (Sept. 1923): 242–58.

3. Davisson and Kunsman, "The Scattering of Low Speed Electrons by Platinum and Magnesium."

4. Carlo Rovelli, "Quantum Weirdness Isn't Weird—If We Accept Objects Don't Exist," *New Scientist*, Mar. 10, 2021, https://www.newscientist.com/article/mg24933250-500-quantum-weirdness-isnt-weird-if-we-accept-objects-dont-exist/.

5. Clinton Davisson, "The Discovery of Electron Waves," Nobel lecture, Dec. 13, 1937, https://www.nobelprize.org/prizes/physics/1937/davisson/lecture/.

6. Beans Velocci, "Wrenching Torque: On Being Professionally Nonbinary," *Historical Studies in the Natural Sciences* 52, no. 3 (June 2022): 476–84, https://doi.org/10.1525/hsns.2022.52.3.476.

7. Velocci, "Wrenching Torque."

8. Personal correspondence.

9. Robert A. Millikan, "The Electron and the Light-Quant from the Experimental Point of View," Nobel lecture, May 23, 1924, https://www.nobelprize.org/prizes/physics/1923/millikan/lecture/; Theodore Arabatzis, *Representing Electrons: A Biographical Approach to Theoretical Entities* (Chicago: University of

Chicago Press, 1965), 71–73; George Paget Thomson, "Electronic Waves," Nobel lecture, June 7, 1938, https://www.nobelprize.org/prizes/physics/1937/thomson/lecture/.

10. Arabatzis, *Representing Electrons*.

11. Joseph J. Thomson, "Carriers of Negative Electricity," Nobel lecture, Dec. 11, 1906, https://www.nobelprize.org/prizes/physics/1906/thomson/lecture/.

12. Thomson, "Carriers of Negative Electricity."

13. Raquel Gutiérrez, *Brow Neon: Essays* (Minneapolis: Coffee House Press, 2022), 14.

14. Claus Jönsson, "Electron Diffraction at Multiple Slits," *American Journal of Physics* 42, no. 1 (Jan. 1974): 4–11, https://doi.org/10.1119/1.1987592.

15. Trans Legislation Tracker, https://translegislation.com/, accessed Nov. 17, 2024.

16. Samantha Riedel, "Republican Candidates Have Spent More Than $65 Million on Anti-trans Ads Since August," *them*, Oct. 10, 2024, https://www.them.us/story/trump-republican-campaign-anti-trans-ads-election.

17. Bostock v. Clayton County, Georgia, 590 US 644 (2020).

18. National Alliance to End Homelessness, "Trans and Gender Non-Conforming Homelessness," https://endhomelessness.org/trans-and-gender-non-conforming-homelessness/, accessed Nov. 17, 2024.

19. Davisson, "The Discovery of Electron Waves."

20. Karen Barad, *Meeting the Universe Halfway: Quantum Physics and the Entanglement of Matter and Meaning* (Durham, NC: Duke University Press, 2007), 99.

21. Paul G. Kwiat, Aephraim M. Steinberg, and Raymond Y. Chao, "Observation of a 'Quantum Eraser': A Revival of Coherence in a Two-Photon Interference Experiment," *Physical Review A* 45, no. 11 (June 1992): 7729–39.

22. Stephan Dürr, T. Nonn, and Gerhard Rempe, "Fringe Visibility and Which-Way Information in an Atom Interferometer," *Physical Review Letters* 81, no. 26 (June 1998): 5705–709.

23. Clara Moskowitz, "Largest Molecules Yet Behave Like Waves in Quantum Double-Slit Experiment," *Live Science*, Mar. 25, 2012, https://www.livescience.com/19268-quantum-double-slit-experiment-largest-molecules.html.

24. Karmela Padavic-Callaghan, "'Dark Photon' Theory of Light Aims to Tear Up a Century of Physics," *New Scientist*, Apr. 29, 2025, https://www.newscientist.com/article/2477695-dark-photon-theory-of-light-aims-to-tear-up-a-century-of-physics/.

25. Melissa Faliveno, *Tomboyland: Essays* (New York: Little A/TOPPLE Books, 2020), 43.

26. Timothy J. Atherton et al., *LGBT Climate in Physics: Building an Inclusive Community* (College Park, PA: American Physical Society, 2016).

27. Justin A. Gutzwa et al., "How Women and Lesbian, Gay, Bisexual, Transgender, and Queer Physics Doctoral Students Navigate Graduate Education: The Roles of Professional Environments and Social Networks," *Physical Review*

Physics Education Research 20, no. 2 (Sept. 2024): 020115, https://journals.aps.org/prper/abstract/10.1103/PhysRevPhysEducRes.20.020115.

28. Chanda Prescod-Weinstein, *The Disordered Cosmos: A Journey into Dark Matter, Spacetime, and Dreams Deferred* (New York: Bold Type Books, 2022), 169.

29. Prescod-Weinstein, *The Disordered Cosmos*, 180.

30. Barad, *Meeting the Universe Halfway*, 140.

31. "Torrey Peters: Stag Dance," *Between the Covers* (podcast), Mar. 24, 2025, https://tinhouse.com/podcast/torrey-peters-stag-dance/.

32. José Esteban Muñoz, *Cruising Utopia: The Then and There of Queer Futurity*, 10th anniversary ed. (New York: NYU Press, 2019), 1; Prescod-Weinstein, *The Disordered Cosmos*, 182.

CHAPTER 11: ELECTROMAGNETICALLY INDUCED TRANSPARENCY

1. Nafis Hasan, *Metastasis: The Rise of the Cancer-Industrial Complex and the Horizons of Care* (Brooklyn, NY: Common Notions, 2025).

2. Steve Alsop, "Affect in Learning Science," in *Encyclopedia of Science Education*, ed. Richard Gunstone (Dordrecht, The Netherlands: Springer Dordrecht, 2015), 24, https://doi.org/10.1007/978-94-007-2150-0.

3. John Tyler, "Degrees Earned in the Physical Sciences and Engineering Fields," American Institute of Physics, Jan. 25, 2024, https://ww2.aip.org/statistics/physics-engineering-degrees-earned#.

4. Barbara Crawford, "Authentic Science," in Gunstone, *Encyclopedia of Science Education*, 24.

5. Joe Feldman, *Grading for Equity: What It Is, Why It Matters, and How It Can Transform Schools and Classrooms* (Thousand Oaks, CA: Corwin, 2018).

6. Lucy Avraamidou, "Science Identity as a Landscape of Becoming: Rethinking Recognition and Emotions Through an Intersectionality Lens," *Cultural Studies of Science Education* 15 (June 2020): 323–45, https://doi.org/10.1007/s11422-019-09954-7.

CHAPTER 12: QUANTUM SPIN

1. awesome123batman, "What Is Spin Exactly? (Quantum Mechanics)," Reddit, June 7, 2020, https://www.reddit.com/r/AskPhysics/comments/gyavoi/what_is_spin_exactlyquantum_mechanics/.

2. Ramamurti Shankar, *Principles of Quantum Mechanics*, 2nd ed. (New York: Plenum Press, 1994), 374.

3. Robert Sintara et al., "A Century of Physics," *Nature Physics* 11 (Oct. 2015): 791–96, https://www.nature.com/articles/nphys3494.

4. David Mermin, *Boojums All the Way Through: Communicating Science in a Prosaic Age* (Cambridge: Cambridge University Press, 1990), xi.

5. Bretislav Friedrich and Dudley Herschbach, "Stern and Gerlach: How a Bad Cigar Helped Reorient Atomic Physics," *Physics Today* 56, no. 12 (Dec. 2003): 53–59.

6. Friedrich and Herschbach, "Stern and Gerlach."

7. Friedrich and Herschbach, "Stern and Gerlach."

8. Friedrich and Herschbach, "Stern and Gerlach."

9. Werner Heisenberg, *Physics and Philosophy: The Revolution in Modern Science* (London: Penguin Classics, 2000), 114–15.

10. See chapters 5 and 10.

11. George Uhlenbeck and Samuel Goudsmit, "Ersetzung der Hypothese vom unmechanischen Zwang durch eine Forderung bezüglich des inneren Verhaltens jedes einzelnen Elektrons," *Naturwissenschaften* 13 (Nov. 1925): 953–54, https://doi.org/10.1007/BF01558878.

12. Jürg Fröhlich, "Spin, or Actually: Spin and Quantum Statistics," in *The Spin: Progress in Mathematical Physics*, ed. B. Duplantier, J. M. Raimond, and V. Rivasseau (Basel, Switzerland: Birkhäuser Basel, 2009), 1–60.

13. Ronald G. J. Fraser, "The Effective Cross Section of the Oriented Hydrogen Atom," *Proceedings of the Royal Society A* 114 (Mar. 1927): 212–21, https://royalsocietypublishing.org/doi/10.1098/rspa.1927.0036.

14. Domenico Giulini, "Electron Spin or 'Classically Non-Describable Two-Valuedness,'" *Studies in History and Philosophy of Science Part B: Studies in History and Philosophy of Modern Physics* 39, no. 3 (Sept. 2008): 557–78, https://doi.org/10.1016/j.shpsb.2008.03.005.

15. Vlatko Vedral, "Quantum Analogies," *My Blog*, June 6, 2022, https://www.vlatkovedral.com/quantum-analogies/.

16. Abraham Pais, "George Uhlenbeck and the Discovery of Electron Spin," *Physics Today* 42, no. 12 (Dec. 1989): 34–40, https://doi.org/10.1063/1.881186.

17. George E. Uhlenbeck, "Fifty Years of Spin: Personal Reminiscences," *Physics Today* 29, no. 6 (June 1976): 43–48, https://doi.org/10.1063/1.3023519.